Welcome to The Book of Making 2

HackSpace magazine is filled with the best projects, tutorials, and articles for makers and hackers. Each year, that amounts to over 1500 pages! Here, we've distilled the second year of HackSpace magazine down to our favourite maker projects. We don't discriminate between different styles of making: in this book we look at how to make vinegar, how we built our first rocket, a clock we made, and when we learned to weld. The common theme through all of these is that they resulted in something physical coming into existence. If, like us, you're interested in building things, then read on: we've

The common theme through all of the the projects featured in this book is that they resulted in something physical coming into being

got some crackers in this book. My personal favourite is the word clock on page 104, but then I'm biased as that now sits in my kitchen. Take a look, and let us know which is your favourite!

BEN EVERARD
Editor   @ ben.everard@raspberrypi.org

Got a comment, question, or thought about HackSpace magazine?

get in touch at
hsmag.cc/hello

GET IN TOUCH

@ hackspace@raspberrypi.org

f hackspacemag

🐦 hackspacemag

ONLINE

hsmag.cc

Features Editor
Andrew Gregory
@ andrew.gregory@raspberrypi.org

Book Production Editor
Phil King

Sub Editors
David Higgs, Nicola King

DESIGN
Critical Media
criticalmedia.co.uk

Head of Design
Lee Allen

Designer
Ty Logan

Photography
Brian O'Halloran,
Fiacre Muller

CONTRIBUTORS
Matt Bradshaw, Andy Clark, Dermot Dobson, Alex Eames, Jo Hinchliffe, Andrew Lewis, Brian Lough, Steve Pelland, John Proudlock, Archie Roques, Mayank Sharma, Barbara Taylor-Harris

PUBLISHING
Publishing Director
Russell Barnes
@ russell@raspberrypi.org

Advertising
Charlie Milligan
@ charlotte.milligan@raspberrypi.org

DISTRIBUTION
Seymour Distribution Ltd
2 East Poultry Ave,
London EC1A 9PT
+44 (0)207 429 4000

SUBSCRIPTIONS
Unit 6, The Enterprise Centre,
Kelvin Lane, Manor Royal,
Crawley, West Sussex,
RH10 9PE

To subscribe
01293 312189
hsmag.cc/subscribe

Subscription queries
hackspace@subscriptionhelpline.co.uk

50 PROJECTS FOR MAKERS & HACKERS

BOOK OF MAKING VOLUME 2

MAKE ELECTRONICS CIRCUITS WITH PLAY-DOH

PLAY GAMES WITH A CUSTOMISED CONTROLLER

TAKE TO THE SKIES WITH YOUR OWN ROCKET

CREATE MUSIC ON A HOMEMADE SYNTHESIZER

FROM THE MAKERS OF HackSpace MAGAZINE

Contents

Rubik's Cube solver

54

Casting LEDs

22

Polyphonic digital synthesizer

88

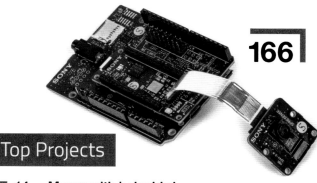

Clothes pegs
114

HackSpace
TECHNOLOGY IN YOUR HANDS

ONE HOUR PROJECTS

HACK | MAKE | BUILD | CREATE

For when you don't have a great deal of spare time, these projects are quick to make – have fun!

HackSpace
TECHNOLOGY IN YOUR HANDS

PG
10

COPPER PIPE LAMP

Steampunk lighting that you can make at home

PG
16

MAKER'S LAB: PLAY-DOUGH CIRCUITS

Make electronic circuits with conductive dough

PG
8

MAKE A CUSTOM RUBBER STAMP

Create beautiful, unique, personalised rubber stamps

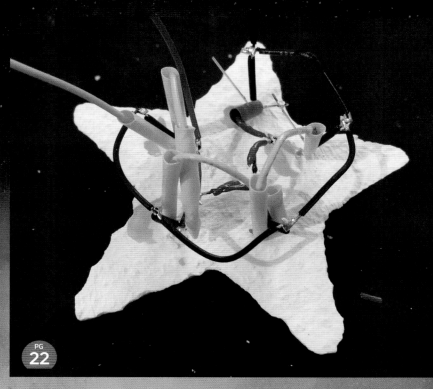

PG
22

Top Projects – Showcase

Mama with baby bird

PG
14

TUTORIAL

Make a custom rubber stamp

Beautiful, unique, personalised rubber stamps in a couple of hours

Alex Eames

🐦 @raspitv

Alex Eames loves making things and regularly blogs/vlogs about this at RasPi.TV He makes a living designing and selling RasP.iO products. raspi.tv

Rubber stamps are great fun and really useful for great for crafting, decorating, authorisation, event 'pass out' verification, and all kinds of short-run print jobs. With widely available access to laser cutters and free design software (Inkscape), you can go from concept to stamping in just a couple of hours.

NEGATIVE THINKING

You're starting with a flat rubber surface and burning away everything except the 'printing area'. So, your image needs to be created in negative form. Everything in black will be 'engraved' away by the laser. Everything in red will be a cut line (used for the stamp outline).

Everything in white will be the rubber 'left behind' that will transfer ink to the paper.

If it's not horizontally symmetrical, your stamp image needs to be flipped horizontally so it will print the right way round. Lettering should be at least 2–3 mm high or it won't be resolved.

Create your image, bearing in mind the above design parameters. Save it as a PNG or JPG file.

IMAGE IS EVERYTHING

Open Inkscape (**inkscape.org**). Select File > Import and choose your image. Leave the import settings at default.

Now define a perimeter cut line for the laser cutter. Click on the rectangle tool (or press **F4**), and drag a rectangle around your design to define the total area of your stamp.

Open the 'fill and stroke' pane (**SHIFT+CTRL+F**). Change the stroke paint colour to red (255, 0, 0) and the stroke style to 'solid 0.2 mm line'. If you want rounded corners on your outline, drag the top-right (circular) drag-handle of the rectangle downwards until satisfied. To fill with black, set the fill colour to black (0, 0, 0). If that makes your graphic disappear, choose Object > 'Lower to bottom' to push this black

rectangle to the background. Then Edit > 'Resize page to selection' (**SHIFT+CTRL+R**). If you haven't already done it, now horizontally flip your image, you can do this by pressing **H**.

You should now have your chosen graphic in white within a black (rounded) rectangle surrounded by a thin red line.

Now save your file at full size and then you'll scale it. To scale your stamp to the exact size you want, Edit > Select All (**CTRL+A**), then lock the aspect ratio of your stamp by ensuring that the little padlock between

> You're starting with a flat rubber surface and **burning away everything except the 'printing area'**

the width and height adjustment boxes is in the 'locked' position (click to toggle it). Change the scale of your stamp to suit your application; e.g. set width to 40 mm. The height will scale automatically. Then, Edit > 'Resize page to selection' (**SHIFT+CTRL+R**) and save your scaled SVG file with a new name (e.g. MyStamp_30x30.SVG) without overwriting the original.

Below ◈
Lettering should be at least 2–3 mm high or it will just be a blurry mess. The largest here is just about acceptable

QUICK TIP

Some laser cutter software uses different colour schemes for engrave and cut lines, so check which you need before you design your artwork.

LET'S BURN RUBBER

Test your laser cutter settings on a small image or a scaled-down version of your stamp to save wasting lots of material. The settings on our K40 are Engrave: 25% power and 200 mm/s, Cut: 25% power at 8 mm/s. Yours will be different, but this gives a starting point for experimentation. Air Assist will help here as the rubber will likely want to flame.

Once engraved and cut out, a blast of air or dabbing with a paintbrush will remove any dust or debris.

Then, using the same cut file, cut a piece of 3 mm acrylic (17% power, 8 mm/s) with the identical outline to your stamp. This will act as a firm backing, providing even pressure across the stamp. Stick the rubber stamp to the acrylic using either a thin layer of glue or double-sided sticky tape.

Then cut a suitably sized wooden handle 2–4 cm long, with cross-section smaller than the acrylic surface area. Sand the handle edges so they are not sharp or rough. Attach one end to the top-side of the acrylic using glue or tape (a hot glue gun is very quick).

Now your stamp is finished and ready to try out. Enjoy! ◻

Above ◈
Once completed, you can start decorating and customising everything. Making multiple copies of a stamp for different colours avoids messy cleaning

Below ◈
The acrylic is stuck to the rubber stamp with double-sided tape. Wooden handle to acrylic is fixed with a hot glue gun

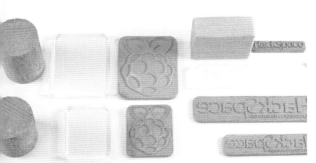

STAMPING TECHNIQUE

Press the stamp firmly (but not too hard) onto the ink pad. Remove and press the stamp onto the paper. On a brand new stamp, it may take a couple of iterations until the ink properly wets the rubber. You'll get better results if you don't press too hard, but pressing firmly with a slight rolling action seems to work well. Practice is good. Be prepared to waste some paper and ink learning the best technique.

YOU'LL NEED

◈ **Computer with Inkscape or Adobe Illustrator**

◈ **Sheet of odour-free laser-cutting rubber**

◈ **Small piece of 3 mm acrylic sheet**

◈ **Small scrap of wood**

◈ **Access to laser cutter**

◈ **Ink pad**

TUTORIAL ━━━━━━━━━━━━━━━━━━━━━━━━━━━━━━━━━

Copper pipe lamp

Steampunk lighting that you can make at home

Ben Everard

🐦 @ben_everard

Ben loves cutting stuff, any stuff. There's no longer a shelf to store these tools on (it's now two shelves), and the door's in danger.

Electric lights are perhaps the most basic electrical item. At their most basic, they contain just two components – a bulb and a switch. This simplicity gives us plenty of scope to put our own artistic twist on the design. We've gone down a common route of using copper pipe to create a custom frame for our design. This pipe is both easy to work with and looks great. A yellow-orange light accentuates the colour of the copper, resulting in a visually pleasing lamp.

When working with mains electricity, we obviously need to make sure that we're kept safe from live current. When building appliances that are hopefully going to be used for many years, we don't just need to make sure that they're safe at the point we build them, but that they'll remain safe. We'll do this by both reducing the chances of something going wrong, and putting a mitigation in place in case it does. We've built a lamp that we're happy with, but ultimately it's your responsibility to understand the risks inherent in working with mains voltage electricity, and make sure your lamp is safe. It's beyond the scope of a magazine article to go through all the risks of working with mains electricity.

The obvious risk with a copper pipe lamp is that, somehow, the live wire gets in contact with the copper pipe. If someone then touched the pipe, they could have a serious shock. Because copper is conductive, it's inherently more risky than making a lamp out of non-conductive material, so we have to be more careful to ensure this doesn't happen.

The lamp could get live in two ways: either the cable could be pulled out of the bulb holder, or the flex could become damaged inside the pipe. To reduce the risk of these happening, we'll make sure that the cable is secured at the point it exits the lamp, so tugging on the cord doesn't put force on the connection between the flex and the bulb holder,

Right ◈
The completed lamp doesn't have to illuminate your wine rack, but it can

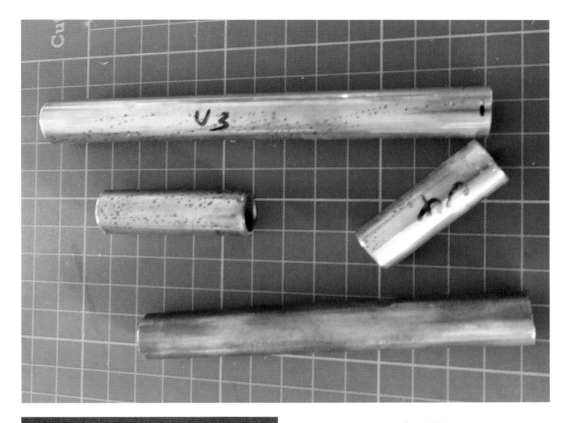

Left ◈
Our copper pipe
ready for assembly.
Getting the cuts as
smooth as possible
will help make
sure everything fits
together properly

SWITCHES AND DIMMERS

Our bulb holder has an integrated switch, but there are a few other options. You can have a torpedo switch on the flex cable (just make sure that your switch has earth pass-through, so the effort to protect your lamp isn't wasted).

Edison-style filament bulbs look particularly good with the copper, and these dim particularly well. You can get in-line dimmers that can be wired into the flex, just as a torpedo switch could be. We find the orange-red light of a dimmed filament bulb very soothing in the evening.

and this will also reduce the movement of the flex inside the lamp, which will also minimise the risk of it being damaged inside the lamp.

To mitigate the risk, should the flex insulation somehow fail, we'll ensure that the whole body of the lamp is earthed. This means that if there is a contact between the live part and the body of the lamp, it'll get harmlessly diverted to ground, and this should either blow a fuse or trip a circuit breaker.

We'll look at how to put these two protections in place as we build the lamp.

SNIP SNIP

Copper pipes are soft and easy to work with. You can get specialist pipe cutters, but most metal saws

will make short work of it. Whichever you use, you need to make sure that the cuts are clean before assembling your lamp. If there are any burrs, they could stop the joints fitting together, or they could dig into the cable, which could have disastrous consequences. You'll need a file that fits inside the pipe, and give it a good scrubbing until there's nothing protruding from the cut.

Let's take a look at our options for joining the pipe together. The most elegant method is to use soldered joints. These are pre-formed joints and bends that slot around the pipe and can be heated and soldered together. Some already contain solder, while others need to have solder added. Whichever version you use, you need to add flux to the pipes first to ensure a good joint. They should fit quite snugly, so you can push the pipes into place and line everything up, then go around with a blow-torch and solder every joint. The big disadvantage of these joints is that you have to seal them without the cable inside, and this makes it hard to thread the cable through more complex shapes. We found that we could thread the flex round a 90 degree bend, but any more than this proved challenging.

The second joint option is compression joints. These work by having three parts: the joint itself that you press the pipes into, then an olive that looks like a ring of brass or copper, and finally a threaded end. These work by screwing the threaded end →

YOU'LL NEED

◈ **15 mm copper pipe** (about 1 metre)

◈ **Earthed bulb holder**

◈ **Assorted pipe joints** (elbow and 'T')

◈ **2 × pipe end stop, with M10 threaded holes**

◈ **Hollow M10 threaded rod** (about 30 mm long)

◈ **M10 cord gripper**

◈ **3 m of flex**

◈ **Plug**

◈ **Blow-torch** (optional)

◈ **In-line dimmer** (optional)

TUTORIAL

Right ◈
A soldered copper
T-joint. The ridges
around each edge
hold the solder
that's heated to seal
the joint

Below ◲
Our bulb holder
with the live and
neutral going into the
fixture, and the earth
secured in the base

onto the joint with the olive between. As the two parts are forced together, there's enough pressure on the olive to deform it and create a solid joint. These are bulkier and this affects the look of the lamp, but they have two advantages: they don't require a blow-torch, and they can be done after the flex is threaded through the lamp, so there's not a problem creating complex shapes.

There are a couple of other options that would join the pipes together but that we can't recommend. There are plastic pressure-fit connectors and you can secure the solderable joints with glue rather than solder. The reason we can't recommend these is because they don't make electrically conductive joints. This means that you can't easily earth the entire metal body of the lamp, so if the live cable frayed in an unearthed part of the lamp, you could have a dangerous object.

You can bend copper pipes to create interesting shapes. The challenge is to create a bend without

the pipe collapsing, and there are a few tools out there for the job, but that's beyond the scope of this article.

GET INSPIRED
You can get creative with the design – the only restriction is that you can actually thread the flex through the design.

We've gone for a three-pointed base, for the simple reason that three legs can't wobble, but four or more can. Copper is a bit forgiving though. If you do opt for more legs and there's a little wobble, you can bend copper slightly by hand to correct any misalignment.

Above this, we've got a vertical pipe with two 90 degree bends that end up hanging the bulb down. Originally, we had both of these corners as soldered joints, but we couldn't thread the flex, so we had to cut one out and replace it with a compression joint. We could have replaced either one, but elected to do the outside one. We've also got a compression T-joint on the upright which gives us a place for our cable to escape.

The exact measurements don't have to be precise – ours is about 30 cm high and 20 cm wide, but we didn't even use a tape-measure, and just cut it by eye where it looked about right. It's useful to have a bulb with you when cutting the lengths out so you can picture how it will look when it's all fitted together.

Typically, bulb holders end in a 10 mm thread, and you'll also need a pipe end stop with a 10 mm threaded hole that you can attach to the end of your lamp. You can join these two fittings with a 10 mm hollow threaded rod, and through this the cable will get from the bulb holder into the body of the lamp.

All these bits you can get from a lamp maker store such as **lampspares.co.uk**.

Make sure that your bulb holder is either metal and earthed (which will earth the whole lamp provided all the joints are conductive), or you find another way to safely earth your lamp body.

At the other end, the flex needs to be secured as it leaves the pipe. Again, we'll use a 10 mm threaded end stop, but this time we'll need a cord gripper with a 10 mm thread. These are most commonly used to secure lamps hanging by their cord from the

> **The exact measurements don't have to be precise** – we didn't even use a tape-measure, and just cut it by eye

ceiling, but also work to secure the flex in this case. An alternative option is to use a rubber grommet to protect the flex, and add a flex strain relief to prevent the cable pulling through this hole. We found that this doesn't fit in a 15 mm pipe, but you might be able to use this method if you're using thicker pipe.

That's all the pieces you'll need – let's put it all together. First, solder any joints you'll be soldering, then thread the flex through all the parts, including compression joints. Join everything together and

wire up the bulb holder and plug, according to the manufacturer's instructions. This could be an easy build or it could be complex – it completely comes down to your design.

We're basically there now, but there's a couple of safety checks before getting started. If you've got a PAT tester, then this is ideal, but if not, we can check some things with a multimeter. First, check that the lamp's properly earthed by ensuring that the resistance between the earth pin on the plug and any exposed metal is low enough (less than a few ohms). Secondly, make sure that the resistance between the live and earth pins is very, very high (this ensures that there's not a leakage already). Remember that, just because your lamp is properly assembled now, it is no guarantee that it will remain that way, and you can repeat this test if you're ever unsure about the safety of the lamp.

That's all there is to it. Your lamp is now ready for use. □

Above ◈
The three parts of a compression joint. In the middle is the olive that is deformed when the joint is secured

Left ◈
Securing the flex as it exits the lamp helps make sure your lamp will continue to work for years to come

Mama with baby bird

By **Kelly Heaton**   ◉ Kellyheatonstudio.com

" **I build functional circuits that seem oddly alive, questioning our definition of machines and offering insight into the 'nature' of life.** It's a challenge to bring art and engineering into balance, but the reward is a form of creativity that mirrors the complexities of our own consciousness: thought and emotion.

"In this free-form electronic sculpture, I used surplus resistors and wire to model a mama bird with her baby in a nest. The mama bird sings thanks to a BEAM circuit, designed by Wilf Rigter, that uses a 74HC14 Hex Schmitt trigger as a complex oscillator. The chirping baby is a classic 'canary sound effect' generator, with an audio transformer and transistor. I've added a MOSFET switch to crudely animate a segmented display, like a squawking beak. Both birds have a photoresistor, by which you can interactively affect their sound. Visit me on Vimeo to see a video." □

Right ◪
The media Kelly used to make this are analogue electronics and wood. To give you sense of scale, the mama bird is the size of a blackbird (there's a video at: hsmag.cc/oxiJcW)

Maker's Lab:
Play-dough circuits

We investigate doughs for electronics

Ben Everard

🐦 @ben_everard

Ben loves cutting stuff, any stuff. There's no longer a shelf to store these tools on (it's now two shelves), and the door's in danger.

Usually, when we build circuits, we use some form of metal as the conductor. This allows electricity to flow, as a quirk of chemistry means that electrons can move freely through metallic structures, and there's little resistance to this flow. However, metals aren't our only choice. One conductor popular with science teachers is conductive dough.

There are recipes for various conductive doughs online but most commercial play-doughs are conductive and can be picked up cheaply. We tested out three brands (Poundland, Nickelodeon Junior, and Play-Doh), and found all performed roughly equally (and within our margin for error).

These doughs don't allow electrons to flow, but instead, the liquid part of the dough contains ions, or charged particles. These particles are usually bound together to form a single substance, but can separate to form an electrical current.

For example, many of these doughs contain salt, or sodium chloride (NaCl). When an electric current passes through salt water, this NaCl splits into Na and Cl ions that carry a positive and negative charge respectively. These ions move through the substance carrying the electric current in the same way that electrons do in a metal.

Let's take a look at what this means in practice.

RESISTANCE IS FUTILE

The most important thing we need to know about our conductors is how conductive they actually are. So, we fashioned our doughs into puck-shaped pieces and used our multimeter to read the resistance of a roughly 1cm-thick chunk. It was about $2\,M\Omega$ (megohms). Obviously we're not going to be able to make a circuit out of this, but there are a few problems with the reading.

Firstly, what chunk of material are we measuring the resistance of? The shape of the blob of dough is critical to the resistance. Obviously, the length is important, as the longer the bit of dough we're trying to pass current through, the more resistance we will have. The width (or more accurately, cross-sectional area) is also really important, as the larger the cross-sectional area, the lower the resistance (you can think of this a bit like having resistors in parallel, or, if you'd prefer, like cars travelling down a multi-lane highway where the more lanes there are, the more cars can travel easily). This also introduces problems because the cross-sectional area is important, but if we're only applying a current to a point on either end, then the full width doesn't come into play.

The second part that introduces complexity is that the multimeter can't measure resistance. It

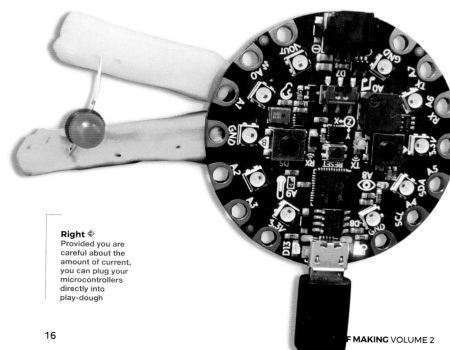

Right ◈
Provided you are careful about the amount of current, you can plug your microcontrollers directly into play-dough

can measure voltage and current – to find out the resistance, it applies a small voltage and measures the current, then uses Ohm's law (I = V / R) to calculate the resistance. This works well for good conductors, but can go astray.

All we want to do is play with some dough and things are getting a bit strange, so let's take a step back and try to simplify it.

We made our dough into pucks about 1 cm deep and 5 cm across, then put a sheet of tin-foil over each end. This tin-foil serves two purposes: firstly it ensures that we get a good electrical connection to the dough across the whole width of the puck, and secondly it protects our electrodes. The ions that arrive from the dough can react with the metal on the electrodes and corrode them. Using foil as a sacrificial electrode protects our real ones.

Now, let's take a bit more control over measuring the resistance. Rather than use the multimeter to measure resistance, we used a bench power supply to provide various voltages, and then used two multimeters to measure the voltage and current, and from this we worked out the resistance of the material at different voltages.

At 0.5 V, our puck had 625 Ω, at 1 V it had 261 Ω, and then over 2 V it stabilised at about 80 Ω. Above about 8 V, it started fizzing and the resistance started to drop again to about 20 Ω at 10 V. You can see the results in **Table 1**.

Because there are ions transporting the current rather than electrons, we end up with an accumulation of different ions at the electrodes. Without knowing the exact composition of the dough, it's a little hard to know what these ions are, but it's a good bet that salt forms quite a sizeable part of the electrolytes in the dough, which means that there'll be an accumulation of chlorine on one electrode.

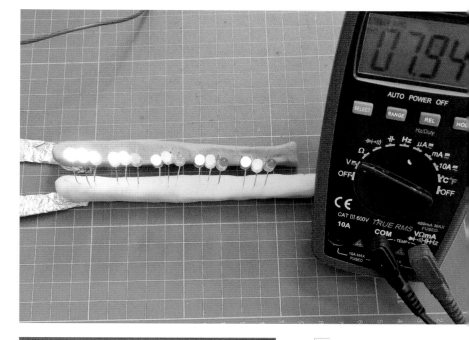

Above ◈
We found that the forward voltage had a big impact on how the LEDs worked, even at voltages well above the forward voltage (red:1.8 V, yellow: 1.86, blue: 2.6)

SQUISHY **CIRCUITS**

We are far from the first people to discover that doughs can be conductive. In fact, there's even a company selling parts specifically for this – Squishy Circuits (**squishycircuits.com**). On the website, you'll find its recipe for dough and a selection of example projects. It also has a store with bits of hardware specifically designed to work with dough.

Some of this is will be absorbed into the water, some of it will react with the various other chemicals there, but some will be given off as a gas (hence the fizzing). This gas isn't a big concern, as long as you keep the current low (as the amount of gas given off is proportional to the current). As a general rule of thumb to stay safe, keep the voltage under 5 V, don't work in a confined space, and don't stick your face too close to the electrode. Don't just keep pumping high voltages through, unless you take proper precautions. →

Table 1 ◈
The resistance of a puck of play-dough at different voltages

VOLTS	CURRENT (mA)	CURRENT (A)	OHMS
0.5	00.8	0.0008	625
1	03.83	0.00383	261
2	24.0	0.0240	83.4
3	33.9	0.0339	88.5
4	56.7	0.0567	70.5
5	63.4	0.0634	78.9
10	423	0.423	23.6

YOU'LL NEED

◈ **Play-dough**

◈ **Power supply or batteries**

◈ **LEDs**

◈ **Multimeter**

◈ **Microcontroller** (optional)

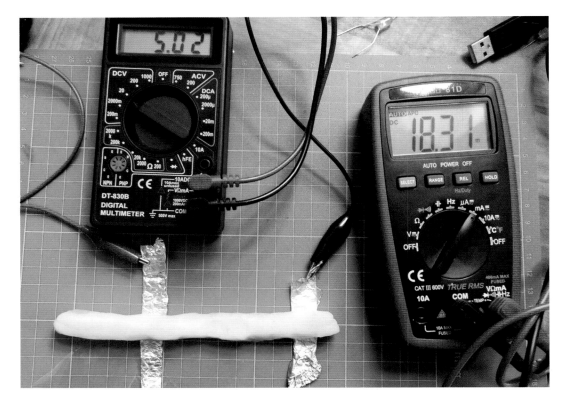

Right ◈
The biggest challenge for testing the resistance was ensuring a good joint between electrode and dough

Table 2 ◈
The resistance of different lengths of a 1cm square cross-section of play-dough

THEORY TO PRACTICE

Let's now take a look at how we can use this dough to build a circuit.

Most red LEDs need a voltage of around 1.8V to turn on. Below this, they simply won't give out any light. The resistance of the dough will mean that we need more than this. In a simple setup with the LED's legs both in a puck of dough, we found that we needed over 2.7V to get a noticeable light on the LED, and 3V gave a reasonable glow.

As we mentioned earlier, the electrodes are subject to quite a bit of corrosion, and the legs of the LEDs are, for these purposes, electrodes. They will corrode and break with use (expect them to last a few hours each).

3V is what we get off most microcontrollers, so you can run this straight from a microcontroller. The biggest challenge is getting the output from the microcontroller into the dough. We used a Circuit Playground Express, since the large pads are easy to press into dough.

Another problem is that it's hard to know what the actual resistance of your circuit is – this is especially true as the connection between your microcontroller and dough might be a significant source of resistance. At 3V, you're unlikely to blow the LED, but more concerningly, you might draw too much current from your microcontroller. Depending on how risk-averse you are, you could try a few options here: using a motor controller, rather than pulling the current directly from the GPIO pin should be safer, as these can handle significantly more current (though check what voltage your motor controller puts out). Alternatively, you could just chance it and, provided you use a long enough length of dough, you should have enough resistance to be safe. How much is long enough, though?

DISTANCE (cm)	VOLTS	CURRENT (mA)	CURRENT (A)	OHMS	OHMS PER CM
1	5	72	0.072	69.4	69.4
2	5	40	0.04	125	62.5
3	5	32	0.032	156	52.1
4	5	26	0.026	192	48.1
5	5	23	0.023	217	43.5
6	5	20	0.020	250	41.7
7	5	18.5	0.0185	270	38.6

Since we got an unusual resistance compared to the voltage, rather than reply on extrapolating the resistance of the dough over length, we tested out the resistance of the dough. We used an approximately 1×1 cm square cross-section (which got a little deformed as we worked) and measured the resistance at different lengths. You can view the results in **Table 2**. As you can see, this is non-linear: at short lengths we had around 70 Ω per centimetre, while at longer lengths we had around 40 Ω per centimetre. This could be because a significant amount of the resistance is at the joint between the electrode and the dough.

On the Circuit Playground Express, the GPIO pins output 3.3 V and are rated to 10 mA (with 7 mA recommended). So, given the 1.8 V forward voltage on the red LED, we need at least 150 Ω of resistance. There will be two lengths of dough (one from the GPIO to the LED, and one from the LED to the ground), so each one needs at least 75 Ω resistance. From our table, we can see that 2 cm should be enough, but prudence suggests a few more, at least

INSULATOR

Perhaps a little surprisingly, it's harder to get an electrical insulator dough than a conductor. Blu Tack is one option, but it can be expensive in large quantities. Fortunately, the good folk at Squishy Circuits (see other boxout) have come up with a recipe involving flour and sugar that has a fairly high resistance, and can be combined with conductive dough to build three-dimensional circuits.

for the initial circuit (it's easy to remove dough, but it's impossible to put the magic smoke back in if you blow your microcontroller).

Remember, all this is based around one microcontroller, and the particular properties of the dough we had. Our experiences with this dough suggest that using 4 or 5 cm of dough for each connection should be plenty to protect most microcontrollers or single-board computers, but don't blame us if yours go pop – this isn't a precise science. ◻

Below ◪
Low voltages proved a particular problem for our play-dough

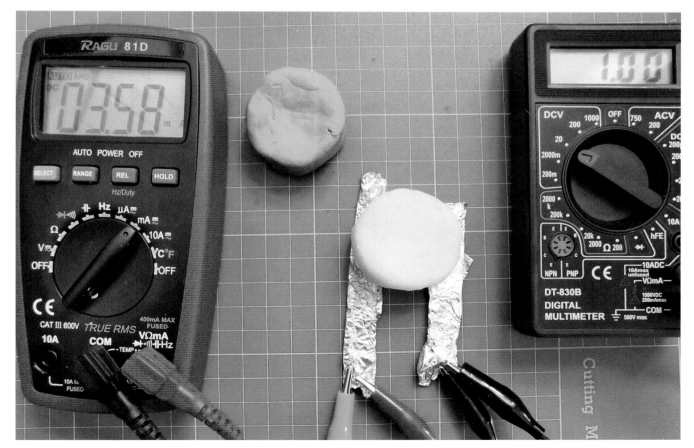

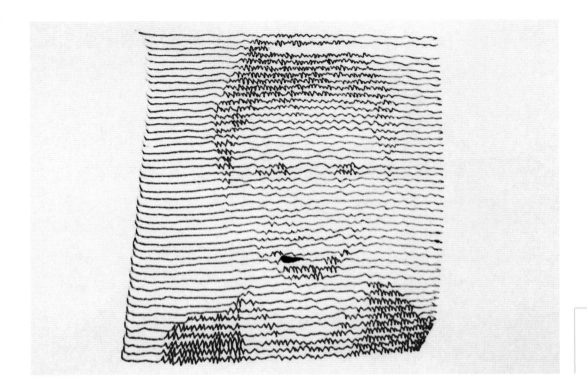

Figure 1 ▨
If you're planning a ghost movie involving polygraphs, this is how you can get your beings to appear

Plotter art

Take mechanical pen to paper and let your computer draw

Ben Everard

🐦 @ben_everard

Ben loves cutting stuff, any stuff. There's no longer a shelf to store these tools on (it's now two shelves), and the door's in danger.

Plotters are devices for moving pens (or other drawing implements) around on paper.** They come in many forms, including classic X-Y plotters (such as the EleksDraw that we review on page 174), and the one we'll be using in this tutorial – the Line-us.

Although we'll be using a Line-us, the same basic techniques should work on most plotters, but you'll have to use different software to control your machine.

We're going to look at two techniques for turning photographs into drawings. You can't directly plot a photograph because it's a raster image – that is, it's made up of a grid of pixels, each of which has a particular colour. Instead, you need an image made up of lines. There's no right way of doing this, and it all comes down to this fuzzy thing they call 'art'. We're going to look at two bits of software for converting images to line drawings, but first, we need to manipulate the image we start with.

We can fit far less detail in a line image than we can in a photo, so we need to make sure we're keeping things simple. We started with a selfie, but we took it against a plain background so there's no extraneous detail in there, then we used our phone's adjustment settings to make our face stand out as much as possible. This involved tweaking the contrast, light, shadows, and highlights settings in a highly unscientific manner until the face was as dark as possible, and the background as light as possible.

The two ways we'll manipulate this into lines are with the SquiggleCam web app and the Stipple Generator. First, let's look at the SquiggleCam. This draws your image in a series of horizontal lines. Each line shows the same frequency wave, but the darker the image at a particular point, the higher the amplitude of the wave (see **Figure 1**). Point your browser to **hsmag.cc/yLgCjQ** and it should load. You can take an image with your webcam, but we found that it worked much better if you adjusted the image as described above.

DRAWING SVGS
ON THE LINE-US

The Line-us doesn't have in-built support for SVGs, but GitHub user, ixd-hof, has made a processing application that can send commands directly to the Line-us from SVG files. You can download this for Windows or Mac from: **hsmag.cc/qYQoXA**. Unzip the file, and run the **Line_Us_SVG.EXE** file. This will open a window with the commands to plot your SVG. Pressing **A** will let you enter the appropriate address for your Line-us (depending on your network, this will either be line-us.local or the IP address of the plotter). Pressing **C** will connect to your plotter (and the line will go green if the connection is successful). **O** lets you choose an SVG to plot. It doesn't automatically scale or centre it, so you need to use +/- and arrow keys to move it to the right place. You'll get more accurate plotting at the bottom of the available range than at the top.

If you load the image and the window stays blank, try zooming out a long way, as it could be in the wrong place.

Once you're ready to plot, hit **P**, and let your Line-us do its thing.

There's a range of things you can change in the Squiggle Settings. The two that had the biggest effect for us were the Line Count, and the Amplitude. The Line Count is the number of horizontal lines. More lines obviously makes for a slower plot, but we found that once we went below about 40, the image was no longer very clear.

Don't be afraid to be heavy-handed with the amplitude setting. The preview image on the SquiggleCam uses a thick line which makes it look more dramatic than it is, and there's a bit of wobble in the Line-us, so the amplitude is dampened a bit when drawn. An amplitude of between 1 and 2 worked best for us.

Once you've got your image how you want it, hit Download SVG, and see box above for how to plot it.

HEAD FULL OF BUBBLES

The second option is the Stipple Generator. This was created by Evil Mad Scientist Labs for their AxiDraw plotter, but it will work with other plotters on account of its outputting SVG files. It converts images into circles using some pretty clever mathematics. The technical details of how it works are detailed on its webpage (**hsmag.cc/qqMQjv**), but if you'd rather just get stuck in, you can download the software at: **hsmag.cc/Mqcbss**. The interface is fairly self-explanatory. The main

configuration option is the number of circles. The default is 2000, but this will take a long time to plot. It's worth dialling this down, at least while you're experimenting.

While the SVG will include perfect circles, this is beyond the Line-us's ability to plot, so instead you get small squiggles. We quite like this effect. The Line-us doesn't manage to make different sized squiggles very well, so it's best to keep the

> Anything that produces a line-based SVG **should work, so get tweaking**

maximum size quite small, and increase the number of circles until you get the effect you want. If you're using a more accurate plotter, you might choose other options.

We've just looked at two options for generating plotter files, but there's far more than this. Anything that produces a line-based SVG should work, so get tweaking, playing with graphics tools, and see what you can find to make your own images unique. Show us what you've done on Twitter **@HackSpaceMag**, and don't forget to include #plottertwitter ◻

Below ◈
Like glitch art and music, the inaccuracies in the Line-us are part of its charm

Casting LEDs for customised displays

Embed blinking lights in whatever shape you like

Ben Everard

🐦 @ben_everard

Ben loves cutting stuff, any stuff. There's no longer a shelf to store these tools on (it's now two shelves), and the door's in danger.

Above ◈
We used crocodile clips to help us test out the casting, before committing it permanently in place

LEDs are little hemispheres, or rectangular beads of light, to add some visual interest to our projects, or feed back data to the user. They're one of the most common electrical components, and we'd be hard-pressed to think of a project we've done that doesn't have at least one of them. However, we don't always want to present a bare LED to the user, so we decided to experiment with casting them inside other materials, to create a more aesthetically pleasing display.

For this project, we used APA106 LEDs, which are WS2812s (also known as NeoPixels) in a through-hole form factor (and they're sometimes sold as through-hole NeoPixels). This means that we get legs that are easy to work with, but can still daisy-chain them together, and control the colours of all of them with a single pin on our microcontroller.

We won't actually wire them up until they are already inside, but it's useful to know how they

connect together, so you can make sure they're in a sensible orientation when you embed them. There are four LEDs: a Positive, a Ground, a Data In, and a Data Out. Just as with regular NeoPixels, we'll connect all the Positives to 5V, all the Grounds to ground, and the Data Out of one LED goes to the Data In on the next. The Data In of the first LED connects to your microcontroller. There are a few different layouts sold as APA106s, so check your vendor's datasheet to see which leg is which.

The official specifications of the APA106 (like most NeoPixel-compatible LEDs) state that they need 4.5–6V in, and the same voltage for the data line. Many microcontrollers have 3.3V out, and you can sometimes get away with using 3.3V in the data connection, if you also use a lower voltage for the power line. However, this is out-of-spec, and prone to random failure. If possible, it's better to use a proper 5V output, such as an Arduino Uno or a 5V level converter. We're particularly fond of the

Above ◈
The LEDs held in place on foam, before
committing them to the casting

Adafruit ItsyBitsy, that is a 3.3 V microcontroller with
one high voltage output, so when connected to a 5 V
power source, pin 5 is a 5 V output. You can also use
a dedicated LED controller, such as a Pixelblaze or
a FadeCandy.

SUBMERSION

Now, that's the electronics hardware sorted. Let's
take a look at the casting. Here, you just need a
mould and something to fill it that will harden. There
are loads of different options for both, and most
should work with this.

> We used plaster because
> we wanted a hidden effect,
> where the user **wouldn't
> expect there to be LEDs**

We won't dive into mould-making here as it's a
deep topic. We made ours with the Mayku vacuum
former, reviewed on page 168. However, you don't
need a vacuum former to do this. You can buy
moulds, or cast them using silicon. You don't even
have to be this complex – if you want a simple
design, you can use anything you have that's the
right shape. Want a rectangular display? Then you
can use an ice cream tub.

The two most common castable materials are
two-part resins and plasters (that you add water to).
Concrete is often used as well, but this is likely to
be too opaque to be useful here. We used plaster
because we wanted a hidden effect, where the user
wouldn't expect there to be LEDs until they light up.
This, however, is up to you.

So far, so straight forward. We've got some LEDs
and something to cast them in, all we have to do

is combine the two. It's complicated slightly by the
fact that we need to hold the LEDs in place until the
material we're casting is solid enough to support
them. Fortunately, our LEDs come with built-in
holders – their legs.

We used a chunk of packaging foam larger than
the mould and pushed the legs of the LEDs into
this foam so it held them in the correct orientation.
Remember that you're holding them in 3D space, so
you might need to shim them underneath to ensure
that they go the correct depth in (but we didn't).
We can now fill up the mould with plaster, resin, (or
whatever you're casting with) and put the foam over
the top – which submerges the LEDs in the plaster.
Now, wait for it to harden, take everything apart,
and you've got your cast light-up material. The only
remaining thing is to solder up the pins on the LEDs,
to make a circuit.

That's all there is to it. This is one area where
the only limit is your imagination, and with a bit
of experimentation, you can build some fantastic-
looking, and surprising projects with a few raw
materials, and a handful of LEDs. ◻

Below ◈
With the LEDs in
place, you can solder
the circuit together
as usual with all the
5 V pins connected
together, all the
Ground pins together,
and the Data Out of
one connected to the
Data In of the next

Earth Clock

By **Simon Robert**   ◉ hsmag.cc/eqtcnn

" **I** **wanted to build something linking electronics, mechanics, and astronomy.** I first thought about making an orrery, but a lot of people had already made a huge amount of them. I wanted to create something that has never been done before, something cool and new.

"I was wondering what I could do, searching for ideas on Instructables, when I saw a contest named 'clocks', I started to think about a clock with an Earth globe on top of it, spinning at the same speed as the Earth in order to see the face of our planet exposed to the sun.

"The realisation was trickier than I thought because there were a lot of parameters to include (the inclination of 23° of the Earth's axis, the difference between a stellar day and a solar day, the modelisation of the sun, finding a way to include the rotation of the Earth around the sun) – turns out astrophysics is hard!" □

Right ▨
The sun is represented by a ring, rather than a single LED, to better simulate the angle that the sun's rays reach Earth

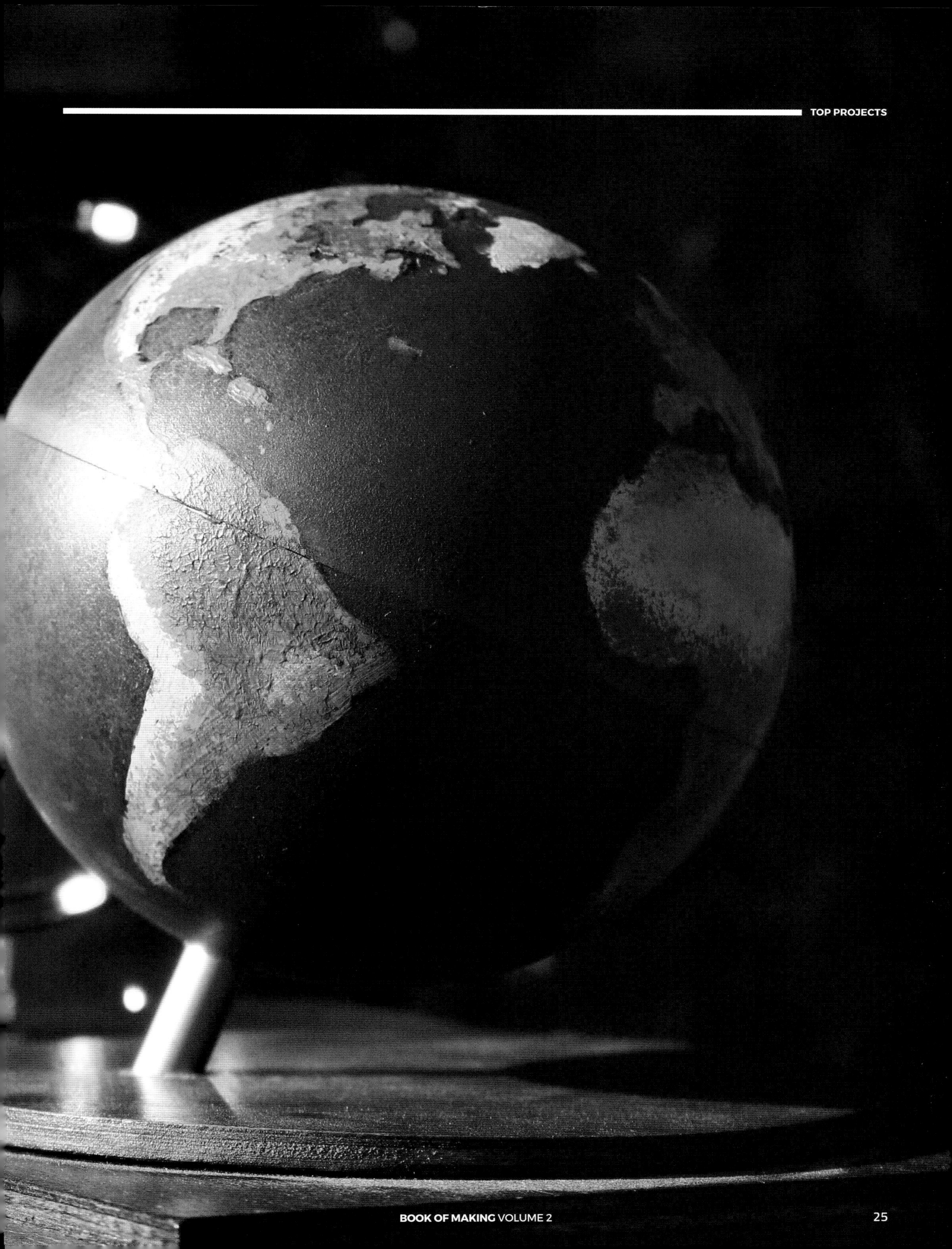

Knife switch

Safely awaken a monster, or just turn on a light, with this remote-control knife switch

Andy Clark

🐦 workshopshed

For the last ten years, Andy has been making and repairing in a shed at the bottom of the garden. You can see more of his exploits at **workshopshed.com**

The knife switch is often seen in classic science-fiction films. But the charm of its simple design is also its major flaw: it can't be safely used to switch on high-voltage circuits. Typically, a mad scientist would pull the lever and sparks would fly as the circuit is connected. We wanted to replicate that but in a safe manner, so we repurposed a Raspberry Pi remote-control socket and added some special effects.

SWITCHGEAR

The knife switch is a double-pole, single-throw switch. It can connect two circuits in parallel. The name comes from the action of the two lever arms that slot between the contacts in a cutting-like action.

The main parts are made from brass, which is a good conductor and easy to work. The connecting handle between the lever arms needs to be an insulator so the two halves of the switch don't join. This part is 3D-printed, and either ABS or PLA should work fine for this part. The base also needs to be an insulator, so we used a 3 mm clear sheet, made of polycarbonate or acrylic.

The arms are made from thicker 2 mm brass strip; this will stop them flexing in use. First, cut them to length. Drill a hole in one end to match your pivot bolt. Drill and tap a hole in the other end for the handle mounting screw. Round the ends with a file.

The hinge contacts require three holes: the first matches the pivots in the arms, the second is used to attach the arms to the case, and the last is for a bolt for the electrical contact. Drill these first, and then mount both in the vice and bend them at 90°.

The contacts are made from a thinner brass strip. Drill the holes for the mounting bolts and contact first. Then, fold the tight bend for the contacts using a nail or rod as a former. Fold the upper contact at 90 degrees. Fold the lower contact, bearing in mind that this fold is in the opposite direction and one thickness further along the bar. Place the upper contact on top of the lower contact, and check that the tops of the two contacts are level.

PLASTIC AND WOOD

The handle is printed from a black plastic, and PLA or ABS should be fine. We used recycled ABS that used to be car dashboards. Printing the cross piece without supports should work fine. The knob is best printed with supports, and sanded smooth. Glue these parts together with suitable plastic adhesive. The backboard for the switch is a piece of clear plastic. Drill the

Right ⬈
Monster control switch

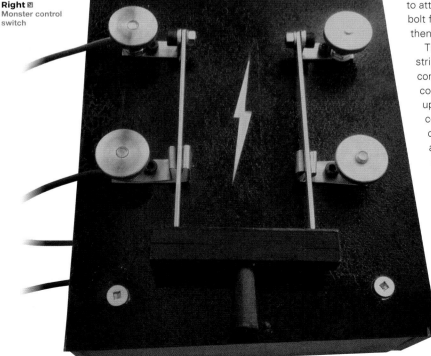

SHIFTING PLATFORMS

Many Raspberry Pi boards can be used without a Pi. Check sites such as **pinout.xyz** to see their connections. Key features to look for are the number of GPIO pins used, voltage levels, and the protocol used to talk to the board. The software for Pi boards is often open source, so you can port it to other platforms. A circuit called a level-shifter can be used if your logic signals are different voltages to the board.

mounting holes for the hinges and contacts, then drill the four corner holes for mounting to the back frame. The rest of the case is made from four strips of wood, with blocks in the corners. Make the top pieces the full width of the plastic, as this is most visible. The side pieces will be the length of the size, minus two times the thickness of the wood you are using. Drill holes in the side for the contact wires.

Use masking tape to mark the electrical spark symbol and to mask the back of these holes. Spray the board with gloss black paint. One or two coats should be sufficient. Once this is dry, remove the tape.

Drill and screw the backboard to the wooden frame.

CODE FOR SPARKS

For the electronics and control software, we can use parts that others have created. The Pi-mote is a small board designed to be used with a Raspberry Pi. It takes a 4-bit encoded signal to turn on one of four remote sockets via a simple radio. Jiri Dohnalek has

 The connecting handle between the lever arms **needs to be an insulator** so the two halves of the switch don't join

produced an Arduino library for this board, so we can use that as a basis of the code. We use an interrupt to trigger the switch being changed.

The first part of the code is to include the library for the Pi-mote. You will need to copy the files from Jiri's GitHub repo – **hsmag.cc/hKgmOs** – into the same folder as your project.

A variable for the socket object is created to allow us to interact with the remote control hardware. We use a variable to store if the interrupt has been

triggered. This variable is set to volatile so that the compiler knows that this variable is changed by the interrupt. We also declare a variable to store the current state of the switch.

```
#include <Pimote.h>
Pimote socket;
byte buttonState = 0;
volatile byte triggered = 0;
```

The setup configures pin 2 as an input, with 3 and 4 as outputs. It also sets up an interrupt which causes code to run when the signal changes on pin 2. This function is called an interrupt handler, or interrupt service routine. The lines for the socket define which pins are to be used and enable the hardware. →

Below ◈
Hinges

YOU'LL NEED

- **Brass strip**
  2 mm and 1 mm
- **Brass nuts and bolts**
- **5 mm socket head screws**
- **Black ABS or PLA filament**
- **Clear plastic sheet**
  3 mm
- **Wood or MDF**
- **Black paint**
- **3.3 V Arduino**
- **Batteries / PSU**
- **LED backlight**
- **100 Ω resistor**
- **Piezo buzzer**
- **Energenie Pi-mote Control**
- **2 × 20-pin Raspberry Pi header**

Above ◩
Leave a gap big enough for the bar

Right ◈
Print separately and glue together

```
void setup()
{
  pinMode(2, INPUT_PULLUP);
  pinMode(3, OUTPUT);
  pinMode(4, OUTPUT);
  attachInterrupt(digitalPinToInterrupt(2),
switchChange, CHANGE);

  socket.setESD0(A0);     // set Encoded Signal D0
  socket.setESD1(A1);     // set Encoder Signal D1
  socket.setESD2(A2);     // set Encoder Signal D2
  socket.setESD3(A3);     // set Encoder Signal D3
  socket.setMODSEL(A4);  // set MODSEL mode select
signal (OOK/FSK)
  socket.setCE(A5); // set CE modular enable
(Output ON/OFF)

  socket.begin();
}
```

The interrupt handler is kept simple, as no other code can be running while the interrupt is run. We set the flag to indicate that the interrupt was triggered.

```
void switchChange() {
  triggered = 1;
}
```

The main loop of the code checks to see if the interrupt was triggered. The short delay gives the switch time to settle. It then activates the LED backlight and buzzer. The state of the switch is toggled and sent to the Pi-mote.

```
void loop()
{
  if (triggered == 1) {
    delay(5);
    buttonState = digitalRead(2);
```

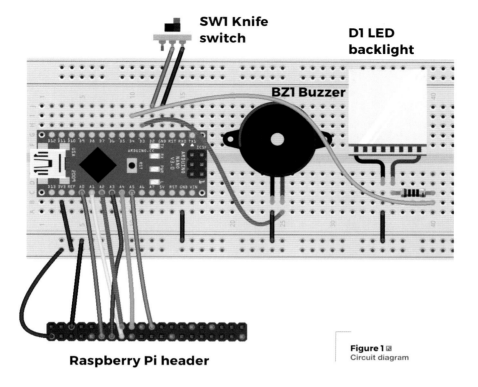

SW1 Knife switch

D1 LED backlight

BZ1 Buzzer

Arduino Nano 3.3 V

Raspberry Pi header

Figure 1 ◪
Circuit diagram

```
    arc();
    if (buttonState == 1) {
        socket.on(1);
    } else {
        socket.off(1);
    }
    triggered = 0;
  }
}
```

> **The electronics for the project consist of** the Arduino, a phone backlight, a buzzer, the Pi-mote board, and the knife switch

The purpose of the **arc** function is to simulate an arcing contact. It does that by flickering the backlight on pin 4 and sending tones to the buzzer on pin 3. The tone frequency is incremented by the square of the loop counter **i** to give a sound similar to a Jacob's ladder.

```
void arc() {
  for (int i = 0; i < 100; i++) {
    int o = random(2); // Random number from 0 to 1
    digitalWrite(4,o);
    float t = ((float)i*(float)i)/1000 + 120;
    tone(3, t ,4);
    delay(5);
  }
  digitalWrite(4,buttonState);
}
```

COMPLETING THE CIRCUIT

The electronics for the project consist of the Arduino, a phone backlight, a buzzer, the Pi-mote board, and

the knife switch. Connect the hinge terminal of the switch to ground, then connect the contacts to pin 2. As the switch is symmetrical, you can connect either side.

Connect one contact for the buzzer to ground and the other to pin 3. The backlight requires a series resistor to limit the current; we used a 100 Ω resistor. Connect this to pin 4 and to the backlight's anode; connect the cathode to ground. Then, connect the pins to the Raspberry Pi header, as in **Figure 1**. ▢

OTHER IDEAS

There are loads of ways you could alter this project to make it your own. A few of our favourites are:

- Use the Arduino sleep modes to put the board to sleep between triggering and improve battery life
- Add a smoke generator
- Add an MP3 of manic laughter
- Create an ultra-modern version with big push-buttons

Making vinegar

Acid makes everything taste better

Ben Everard

🐦 @ben_everard

Ben loves cutting stuff, any stuff. There's no longer a shelf to store these tools on (it's now two shelves), and the door's in danger.

Right ◈
A bottle of our finished vinegar, ready for drizzling on cod and chips

Below ◈
You should end up with a pH around 3 or 4. However, it's better to be guided by taste than numbers on a strip

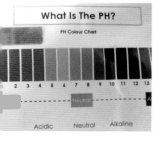

Vinegar is one of those mystical household substances. It's there. We use loads of it. If you leave a bottle of wine open too long, it starts to taste like it, but where does it all come from? It's actually really easy to make high-quality vinegar, and it's far cheaper to make than to buy when compared to commercial raw vinegars.

The starting point for vinegar is ethanol – plain old alcohol that's found in beer (which makes malt vinegar), wine (which makes wine vinegar) and whiskey (which is sacrilege to turn into vinegar). You can make vinegar from alcohol you've bought, but here in the UK (as in many countries) the tax policy makes it expensive to buy alcohol to turn into vinegar, so let's go back a step. The raw ingredient for alcohol is sugar-water (or at least, some liquid with both sugar and water in it), and one of the easiest to work with is apple juice. Fermented all the way through, this produces apple cider vinegar.

Regular supermarket apple juice – the stuff that comes in one-litre cartons – works perfectly well for this. It's fine to use cartons of apple juice from concentrate, but make sure that it's only apple juice (a little added citric or ascorbic acid is fine) – some brands of 'apple juice drink' contain a whole host of other additives, including sugar and other fruit juices. These might be fine or they might end up tasting strange.

In principle, you can use any fruit juice for your vinegar making, but it will affect the end flavours for better or worse – it's up to you how experimental you feel. Most commercial fruit vinegars, such as raspberry, aren't fermented from these fruits, instead they're regular vinegars that are then combined with these fruits after fermenting.

Whatever juice you start with, the key thing you need to check is the amount of sugar in it. The amount of sugar will affect the amount of alcohol, and this will affect the amount of acetic acid in the vinegar. The amount of acid you want depends on what you want it for. Regular table

vinegar is usually about 5% acid, while pickling vinegars can be 10%. As a rule of thumb, you'll end up with 20% less acid than the amount of starting alcohol (it depends a bit on the particular microflora that gets to work in the ferment).

SWEETEN THE DEAL

Approximately 20 grams of sugar per litre will equate to 1% alcohol. We wanted our ferment to end up around 8% acid, so we wanted about 10% ethanol, so we needed 200 g of sugar per litre. Our apple juice had 114 grams of sugar per litre, so we needed to add an additional 85 grams per litre of ordinary caster sugar. All these calculations are subject to quite a bit of error, but they give us an indication of where to start.

The only additional things we need are yeast and yeast nutrient (in principle, you don't need yeast nutrient for cider, as the apples should contain all the nutrients the yeast need, but we add some to help ensure a clean ferment). We used sparkling wine yeast which is available as freeze-dried pellets in a sachet, but almost any brewing yeast should work.

That's the ingredients, now to the equipment. For the first ferment (sugar to ethanol), you need an oxygen-free environment. We used a glass five-litre glass demijohn (or carboy) fitted with a brewer's airlock, but anything that you can fit an airlock on should be fine.

 We used sparkling wine yeast which is available as freeze-dried pellets in a sachet

You need to make sure all this equipment is clean, but we don't sterilise our brewing equipment. Some authorities on home-brewing recommend it, but after several years of brewing without losing a batch, we're no longer convinced it's necessary, provided you're brewing to a high percentage and with an acidic liquid (as apple juice is, even before it's fermented). Beer brewers with the lower alcohol and acidic content may find sterility more important at this point.

We made four litres. First, add three litres of apple juice to the brewing container, then add all the sugar, put your hand over the demijohn opening and shake. This does two things – it helps dissolve the sugar, and it helps bring oxygen into the fruit juice. →

HAVING A DRINK

We started by making cider that's quite similar to regular drinking cider, however, we haven't designed this to be a good drink. The apples used in commercial apple juice are dessert apples, and lack the acidity and tannins needed to make a good drinking cider. In vinegar, this doesn't matter because the acid level is strong enough to give it the body and character that it will lack before the second ferment.

If you like the style of West Country scrumpy ciders, it's a quite nice drink a couple of days into the aerated second ferment. It's got the acidic tang of scrumpy, but not the funk. However, if you plan to drink it this way, you might want to tone down the sugar to make it a more manageable percentage.

Below ↙
During the first few days of the first ferment, you'll get a thick, brown foam on top of the liquid. This is normal

Although you want an oxygen-free environment for the main part of the ferment, having oxygen at the start helps the yeast colonise the liquid. Really, you should start with an open container with more airflow than a demijohn has. However, we've found that as long as you give it a good shake, you get enough oxygen to start your ferment.

You don't need to shake until all the sugar is dissolved, but once most of it has gone, top up with the remaining litre of apple juice. Mix the yeast with a small amount of lukewarm water and a pinch of sugar. It should start to froth and smell like baking bread within about 20 minutes. Once it does, add this to the apple juice and pop the airlock on top.

You'll need to keep the demijohn between about 18 and 33 °C. Much lower than this and the ferment will stall. Much higher and you will kill the yeast. Around 20–24 °C is ideal.

In a couple of days, the liquid should start to froth with a brown foam. If there's no signs of life (bubbles) in the first three days, take the airlock out and give it a good shake. If there's still nothing a couple of days later, shake again, and add more yeast.

About a week later, the brown foam should disappear and it will be replaced by a much smaller covering of white bubbles. It'll keep happily fermenting like this for a couple of weeks depending on temperature – in cold weather it'll take longer. Once it's finished bubbling, you've got cider (or hard cider as it's known in America). If you drink alcohol, take a taste to make sure it's no longer sweet.

A SECOND FERMENT

The next step is to turn this cider into vinegar, which requires acetic acid bacteria. Unlike the yeast, these require an oxygen-rich environment all the way through the ferment. We achieved this using an aquarium pump and (clean) air-stone, but you can also get there by just leaving the liquid open to the

Right ◈
During the second ferment, secure a cloth over the top of the fermenting vessel to protect the vinegar from insects

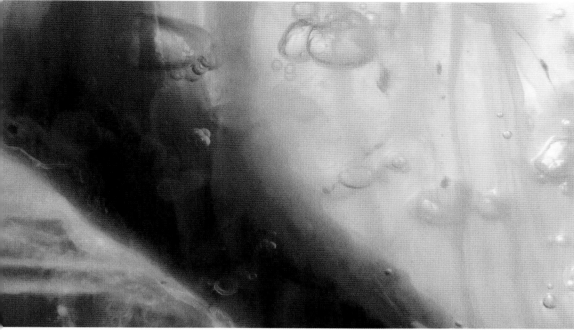

Left ◈
The acetic acid
bacteria form mats
around the air-stone
that trap bubbles

USING YOUR VINEGAR

Your four litres of vinegar will last almost indefinitely if kept in an airtight container, so you don't need to rush to use it all, but here are a few of our favourite ways of using vinegar:

Pickling: Mix your vinegar with a few spices, and perhaps some sugar, and pour over onions, cabbage, chillies, or any other vegetable for a pickled delight. If you plan on storing your pickles, you need to make sure that the pH is low enough to protect the food.

Salad dressings: Mix your vinegar with a good, flavoured oil, such as olive oil, (three parts oil to one part vinegar) and add a little salt and pepper for a classic vinaigrette.

Drinking: Dilute, to taste, with water for a refreshing sour drink.

Marinades: Combine with herbs or spices, and rub onto meat. The acid helps tenderise the meat, ensuring the flavour of the herbs and spices penetrate further into the meat.

air if you're willing to wait long enough (it might take six months or so). Whichever option you choose, you'll need to protect the liquid from flies as these can lay maggots in the proto-vinegar. The advantage of pumping air in is that you only need a small opening in the top of your container, and this is easier to protect. We continued to use the same demijohn that we'd done the first ferment in. It's best to make sure that whatever container you use is safe in acidic environments. Metals tend to corrode and plastics can leach chemicals, so glass is a good choice.

> **Acetic acid bacteria join together** to form colonies

At this point, there should be a build-up on the base of the demijohn. This is dead yeast known as 'lees'. It's best to remove this, as it can lead to off flavours. This just means you need to pour the cider out into a bowl leaving the last centimetre or two, then rinse out the demijohn, and pour the cider back in.

ADDING BACTERIA
This also has the secondary effect of freeing up a bit of space at the top of the demijohn. We need to introduce acetic acid bacteria. These are fairly common, so if you just proceed without adding anything, there's a good chance it will work, but to

speed everything up, and reduce the chances of something going wrong, it's a good idea to add some raw cider vinegar 'with the mother' – you can buy this from health food shops or larger supermarkets. The mother is a common term for the live acetic acid bacteria that we want. There's no hard and fast rules for how much to add – some sources recommend 20%, but that's expensive for your first batch, and you can get away with far less. We use about 100 ml per four litres, but if you've got more, add it.

Pop the air-stone in and cover with cloth (we used a flannel). You'll need to secure this quite well with a rubber band or string to prevent anything getting into the vinegar that shouldn't.

Once the air-stone is bubbling, leave it to ferment. Acetic acid bacteria join together to form colonies. These can look a little dramatic as they grow off the air-stone or float on the surface. They can have the appearance of a cross between pondweeds and aliens, but it's a natural part of vinegar fermenting. Taste once a week or so – it should start to taste a bit vinegary after a few days, but might take up to a month to go all the way. Once it's fermented, decant into bottles and discard this growth (technically, this is the mother, but you never see it in bottles of vinegar sold 'with the mother').

The final result can be very powerful – you can water it down, or temper the acidity with a little sugar – but behind the sharpness, it should be fruity and delicious. You've now got four litres of raw, unfiltered, delicious cider vinegar to pickle, marinade, drizzle, and generally enjoy. How will you consume yours?

Delicious rotten food

Using bacteria to add flavour and keep you healthy

Ben Everard

🐦 @ben_everard

Ben loves tiny things like bacteria and 0201 LEDs. Being small, more of them fit into Ben's makes, and that makes him happy.

Most of the time when we think about bacteria in food, we think about eliminating it. We cook and refrigerate specifically in order to keep our food germ-free (or at least with as few germs as possible). However, this isn't always necessary or desirable. After all, our bodies contain more bacteria cells than they do human cells, and most of these bacteria live in the gut.

These 'friendly' bacteria are an essential part of our health that we're only just discovering the importance of. They help us digest food, produce nutrients, and fend off other bacterial species that could make us ill. As well as this, they can help make delicious food. Yoghurt, cheese, and sourdough bread are just a few foods that get much of their flavour from the bacteria that they contain. However, we're going to look at one of the most iconic bacterial foods available – sauerkraut.

Traditionally, sauerkraut is fermented cabbage (sometimes with carrot added). However, the actual vegetable doesn't affect the fermentation. We'll stick with the traditional cabbage, as it's reliable and delicious, but you can use almost anything if you're feeling experimental (though some end up with an unpleasant texture).

The basic idea is that we keep the cabbage submerged in slightly salty, acidic liquid and this liquid stops any undesirable bacteria from growing. However, we don't need to add any acid – the starting conditions (slightly salty and underwater) will encourage lactic acid bacteria to thrive, and it's these bacteria that will convert sugars in the vegetables into acid. We don't need to add the bacteria – they exist naturally on almost every surface, so they'll be on the cabbage and your hands before you start. In other words, we let the lactic acid bacteria do the hard work of protecting and flavouring our cabbage for us if we give them the right conditions to grow.

There are only two key variables in a sauerkraut recipe: salt and water. There's no fixed amount

Right ◈
The finished sauerkraut. With all the cabbage underwater, it's protected from mould, and the lactic acid bacteria can perform their magic

required for either. Salt does a few things: it's a flavour enhancer, it helps keep the vegetables crunchy, and it helps preserve the ferment.

Traditional recipes include quite a lot of salt because it used to be important for the ferment to last for a long time. However, the acid produced by the bacteria can be enough to protect the sauerkraut for a moderate amount of time. If you're on a low sodium diet, it is possible to omit salt entirely, but you'll need to be particularly vigilant to avoid surface

 There are only two key variables in a sauerkraut recipe: salt and water

moulds. A little salt will make your ferment more likely to be a delicious success. Remember that it's easier to add salt than it is to remove it, so start with a little (about 1.5 tsp per 500 g of cabbage), and add more if you think it needs it for flavour.

FERMENTATION STATION
Once you've decided on your amount of salt, it's time to create your kraut (we'll look at water later). Start by chopping your cabbage – any variety will work. We've used Savoy, because that's what the greengrocer had in stock on the day we did this. Chopping the cabbage finely can make it easier to eat the final product, but it will ferment in any size (and there are some traditional recipes for fermented whole-heads of cabbage). Similarly, add any other vegetables you like at this point. You can also add spices if you wish. Chilli and ginger are delicious additions, and caraway is another popular addition, but you can be as creative as you like.

It's important to start with clean equipment, but there's no need to sterilise it – after all, we are relying on it picking up friendly bacteria, and these will out-compete any unhealthy bacteria that happen to get in.

Put the shredded vegetables and the salt in a bowl, and massage them together. You should find that juice starts to come out of the vegetables at this point.

You'll need a pot for your kraut to ferment in. This should have an open neck that you can easily fit the vegetables in, and you'll need to compact the vegetables later. It shouldn't be metal (which can corrode with the salt) or plastic (which can leach →

Above
We chopped our cabbage into quarters, then finely sliced each quarter, but chop your veg in any way you want

SAFETY
As we're talking about eating food covered in bacteria, you might be expecting a lengthy safety section where we detail arcane protocols to follow to ensure that things remain food-safe. However, no such section is needed as bacterial ferments are incredibly safe. The reason for this is that the bacteria do the hard work of protecting the vegetables for us. The lactic acid bacteria that we want to grow will naturally create conditions that other bacteria can't survive in. This process of protecting food is thousands of years old and predates modern hygiene.

That said, it is important to use your common sense. If something seems wrong – either in flavour, smell, or look – then it's best to err on the side of caution and throw the batch away and start again. It shouldn't smell or taste rotten – the smell should be 'cabbagey' and the taste should be tart and tangy. The colour shouldn't change, unless you add something like red cabbage or beetroot which will spread through the rest of the ferment.

The most common problem is mould appearing on the surface. This furry stuff is due to fungi (not bacteria) and will only grow where it's got access to oxygen (so on top of the water, not underneath it). This is one of the reasons why it's important to ensure that all the cabbage is underwater. If there's just a little furry bit, you can remove the mouldy bits and leave the sauerkraut to keep fermenting. However, if you find a large patch of fuzz, it's probably safer to ditch the lot and start again.

hand in the jar, you'll have to improvise with whatever you have available – rolling pins work well.

Once you've got all the vegetables in the jar, you'll need something to weigh them down. This needs to hold all the veg underwater. The traditional method is to lay a cabbage leaf across the surface and then add a clean stone on top of this; however, any way of holding the veg underwater will work. If you've got a jug or other pot that fits inside your larger pot, this is a great option (and a little water in it will help weigh the veg down further).

At this point, you'll probably find that there's not enough water to completely submerge your vegetables. Don't worry – a little more liquid will leach out of the cabbage over the next few hours, but if there's any veg still exposed after 24 hours, it's best to top up with a little water. Some people recommend using water without chlorine (as this could kill the bacteria), but we've found that it's worked fine with regular tap water. You shouldn't need to add much, so it's unlikely that there'll be enough chlorine to cause any problems – however, if you find yourself adding a lot, or if you live in an area with a high chlorine level, you might want to either use mineral water or take steps to remove some of the chlorine first.

chemicals in prolonged contact with acid). Ceramic crocks are traditional, but glass jars work just as well.

You now need to layer your vegetables into the jar. Put them in a bit at a time, and press them down. Packing them down like this will help keep them submerged under the liquid. It's easiest to put them in a handful at a time and then use your fist to pack them in as tightly as possible, but if you can't fit your

Below ◈
Massaging the salt into the cabbage starts the process of drawing liquid out of the vegetables

Once the cabbage is fully submerged, wipe away any that's stuck to the side of the jar above the waterline, as this will encourage mould growth. Now you just need to cover the ferment with a towel, or other cloth (to keep flies away). Keep it at about the right temperature and wait. Ideally, it should be kept

 There's no point where the kraut becomes finished ... **it'll continue to get more tangy with time**

at 20–23 °C – about room temperature in the UK. A little outside this range shouldn't cause problems, but might affect how long it takes (with it fermenting quicker the warmer it is). If it's below 15 °C, then you might find that it doesn't ferment.

There's no point at which the kraut becomes finished. After a few days, it'll start to develop its distinctive tang, and it'll continue to get more tangy

with time. Typically, it's eaten between a few weeks to a month (or two in cold weather). If you want to keep it longer, you can stop it getting more tangy by putting it in a fridge, or a cool area such as a pantry or cellar. You may want to put a top on the jar at this point, but be aware that it might continue to create gas so you'll need to periodically open it to let out the pressure.

Your sauerkraut is now full of both flavour and beneficial bacteria. You can enjoy it as a condiment, or as an addition to soups and stews. □

OTHER FERMENTED
VEGETABLES

Sauerkraut (or sometimes kraut) is commonly used as a catch-all term for shredded vegetables fermented underwater. It's endlessly varied by using different vegetables and other flavourings, but there are other types of bacterial ferments. Kimchi is a popular Korean ferment, typically made with cabbage, radish, and a range of spices. It's made by a similar process to sauerkraut, but with less liquid. It relies on other methods to keep the oxygen out (often a slightly sealed jar).

Dill pickles are traditionally fermented (though some modern methods use an acid rather than acid-generating bacteria). They're submerged whole in water with herbs and spices, and left to ferment.

Ginger beer is traditionally fermented using a combination of yeast and bacteria, which combine to give it a tang and effervescence. While many modern recipes are simply carbonated ginger-flavoured water, you can still get the traditional cultures to ferment it (called a ginger beer plant, though it's not a plant), or use a ginger bug, similar to the way we've created sauerkraut.

Water kefir uses a combination of yeast and bacteria (similar to traditional ginger beer) to ferment sugars, and usually some fruit, to make a fizzy drink with a slightly sour tang.

Below ◈
Kimchi offers another approach to fermenting cabbage

AFTERNOON PROJECTS

HACK | MAKE | BUILD | CREATE

Got an afternoon to spare? Try one of these projects to learn something new or just have fun tinkering

Top Projects – Showcase

Rubik's Cube solver

PG **54**

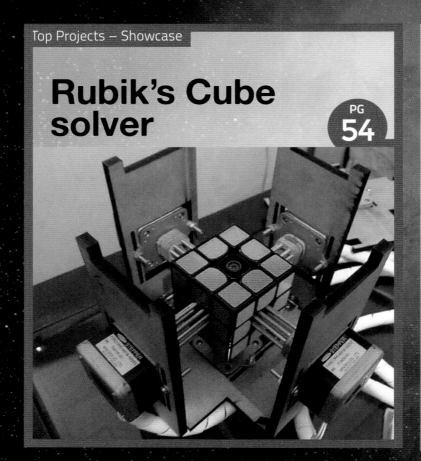

PG **40**

MAKING A 3D PEN DRAGON

Taking sketches to the third dimension

PG **46**

CONTROL A K40 LASER CUTTER OVER A NETWORK

The easy way to send jobs to your cutter

MONKEY SKULL TREASURE BOWL

Anyone reaching into this bowl will be in for a big surprise

PG **62**

TUTORIAL

Making a 3D pen dragon

Taking sketches to the third dimension

Barbara Taylor-Harris

🐦 @BBarbarartist

Barbara is a sculptor, mixed media painter and 3D pen specialist. She's the author of *Go Beyond Doodling*.

A lthough the 3D pen was introduced as a toy, it has become a useful drawing, design, and construction tool. With imagination and practice, it is possible to design and make almost anything, and anyone can learn to use a 3D pen.

The basic techniques that we'll go through here are: drawing lines, simple infill methods and some basic joins. With these, you can design and sketch your own 3D objects.

Working with a 3D pen is similar to when you learn to draw and write with crayons, pencils, and pens. Your first few steps need to be learning how to use your 3D pen by practising drawing lines and gaining control. You are drawing with liquid plastic. It take a little while, but practise will pay dividends.

Basic top tips:

- Always draw away from the molten plastic
- Draw slowly and steadily
- Stop, and then wait before removing the pen

Set your pen to the slowest speed. Drawing slowly at a slow speed should give you a reasonably thick, strong line. The trick is to master matching your drawing speed to the extrusion speed of the pen. You are aiming to create even and consistent lines.

LINE PRACTICE

1. Anchor and draw away
Always press your pen down to your drawing surface. Hold the pen in one

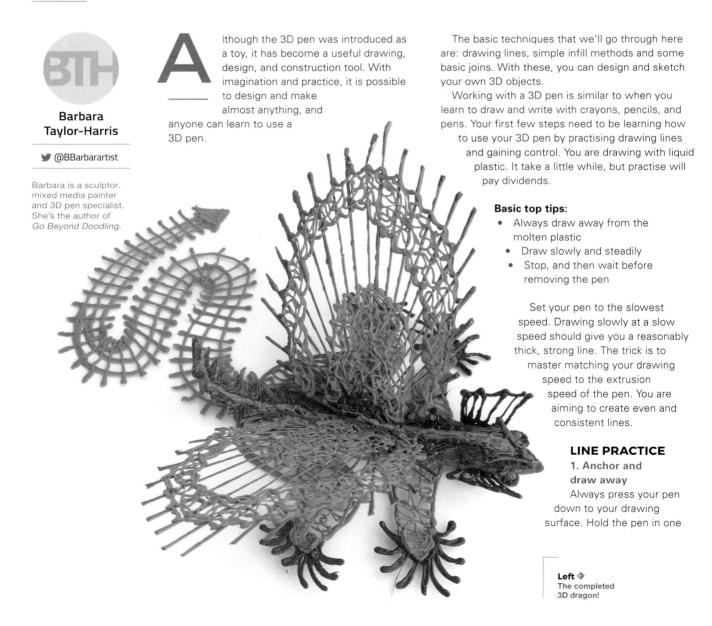

Left ◈
The completed 3D dragon!

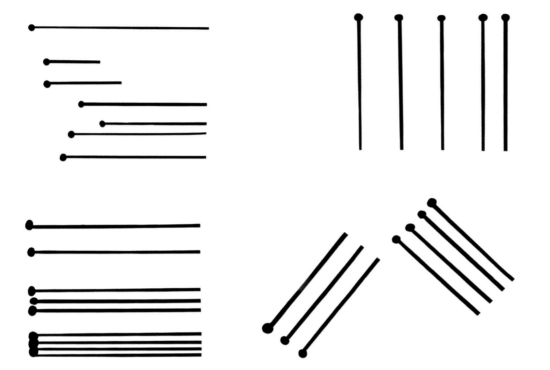

YOU'LL NEED

◈ **3D pen**

◈ **Filament**

place to extrude a small blob to anchor the strand. This will hold the plastic and enable you to draw away slowly.

2. Straight lines

Follow the short lines horizontally, tracing the pen line and using the template lines as a guide to help you. Do this a few times and then turn the template and try drawing vertically and diagonally. Try with and without the metal ruler.

DRAWING THE DRAGON'S WINGS

Draw the outline and then the veins of the wings. Use the metal ruler to guide your pen on the long lines. Draw along the edge of the ruler as you would with any pen or pencil.

To give the wings more strength, two sections use the filigree or lace fill you used on the spine decoration.

This time try drawing with long continuous lines, lifting slightly when you cross another line. Adding more lines and fewer gaps will make the wings stronger but also denser.

When you remove the wings from the tracing paper you will have the drawn side and a smooth side; I usually put the drawn side to the inside when I attach the wings to the dragon's body. →

BASIC **EQUIPMENT**

A 3D pen
There are over 250 pens on the market. Starting off, I advise you to choose one at a lower price to try and see if you like it before spending more.

Some filament
To begin, I suggest you use PLA filament as it has no fumes or odour, and is biodegradable. About 20 metres of one colour.

Good quality tracing paper
You need this over the template to protect it so you can use it several times.

A metal ruler
This will help you draw straight lines. Foam-backed rulers are useful because they give you a little height above the template.

A board/mat
You will need something heatproof to protect your work surface.

Sharp scissors or cutters
You always need to cut a straight end to the filament before feeding it into the 3D pen.

Left ◹
Use these practise lines to get to grips with your pen

TUTORIAL ━━━━━━━━━━━━━━━━━━━━━━━━━━━━━━

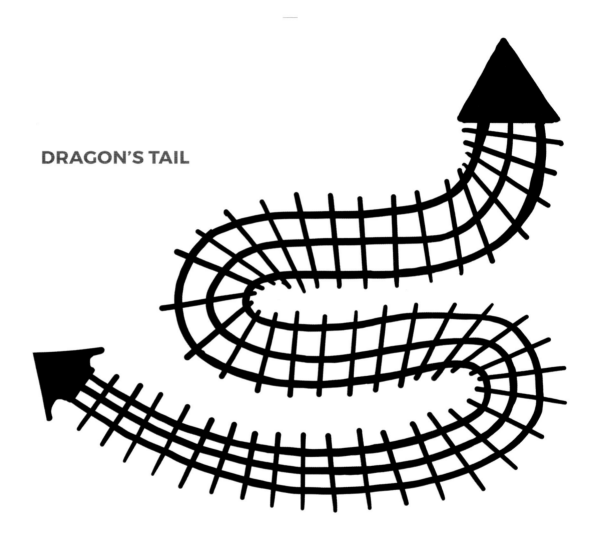

DRAGON'S TAIL

Above ▣
The cross-hatch of
the tail gives both
texture and rigidity

DRAWING THE DRAGON'S TAIL

All lines must join for the plastic to fuse together.
Draw the outlines and the centre lines of the
tail first.

Start at the tip of the tail and work downwards,
then go back and add the crossing lines.

You can make thicker lines and completely fill the
top and bottom of the tail by drawing lines touching
each other. The spaces between the lines on the tail
will help to make it flexible.

COLOURS

One great thing about 3D pens is that it's quick to
change colours. Just swap over the filament and pick
up where you left off. We did the dragon in green and
red, but you can use whatever colours you like. Get
creative! You don't have to limit yourself to colours:
you can also get metallic and glow-in-the-dark
options. The only limit is your imagination (and the
laws of physics, but mostly your imagination).

DRAGON'S SPINE

DRAGON'S WINGS

Left ◈
The spine is a dramatic feature on the top, so we've used plenty of plastic

Below ⬉
The gaps in the wing help give the impression of light, aerodynamic structures

DRAWING THE DRAGON'S SPINE DECORATION

Draw the outline first. Then draw the inner edge of the solid area. Leave the middle blank and carefully fill in the outer shape by making small circular motions with your pen set to slow speed or fill the outer shape by using lots of small lines close together.

If you have gaps you can repeat the process and fill the gaps afterwards.

To fill the middle section and give some rigidity to the shape, we use a filigree or lace infill technique. These lines join the sections together to give the shape strength.

The key is to anchor to one edge and then follow the line to the other side, ensuring all the lines touch and fuse together. Alternatively you could bridge across the gap with straight lines.

DRAWING THE DRAGON'S WINGS

You'll need two of these. Draw the outline and then the veins of the wings. Use the metal ruler to guide your pen on the long lines. Draw along the edge of the ruler, as you would with any pen or pencil.

To give the wings more strength, use the filigree or lace fill you used on the spine decoration.

This time try drawing with long continuous lines, lifting slightly when you cross another line. Adding more lines and fewer gaps will make the wings stronger, but also more dense.

When you remove the wings from the tracing paper you will have the drawn side and a smooth side, I usually put the drawn side to the inside when I attach the wings to the dragon's body. →

DRAGON'S
BODY

Above
You could easily
adapt this design to
be any other repitle

DRAWING THE DRAGON'S BODY

The pattern on the dragon's body uses lines and filigree infill. Draw the outline first, then the internal solid lines. Finally, add the infill. Remember to be sure that all lines join to the outline and/or each other.

ASSEMBLING YOUR DRAGON

First, attach the tail to the top of the body. To do this we use a technique called spot welding. Hold your pen in one place where you want to attach the tail and allow some plastic to puddle. Before it has time to set, position, and press down the top of the tail. Hold down until the filament has set.

Now turn the body over and add some more plastic to securely join the two pieces together.

Next, attach the spine decoration to the body using the same spot welding techniques, and reinforcing on the underside. Then take your wings and position on the body and secure with spots of plastic. You need to hold them in place until completely set.

Then, carefully run a line of plastic along the inside and outside of the joins to strengthen then. Remember to hold the wing in position until both sides of the new joining plastic are set. Ensure there is a secure join and repeat if necessary.

You can use your dragon as a flat creature and attach to a box lid for decoration, or simply hand it around. But it is possible to give it a more 3D appearance. Open filled shapes can be gently bent. Try bending the feet outwards. If your plastic will not bend easily, simply run a line of new plastic along the thick line between the foot and the leg. The new hot plastic will heat and soften the other plastic around it. When it is cool enough to touch but not set, you will find it should bend easily. Do the same where the front legs join the body, and bend inwards. On the back legs make two bend points on the legs. You may need to keep bending and shaping to make your dragon stand. Gently bend the wings and the tail to add a little movement to them. Bend the gills upwards.

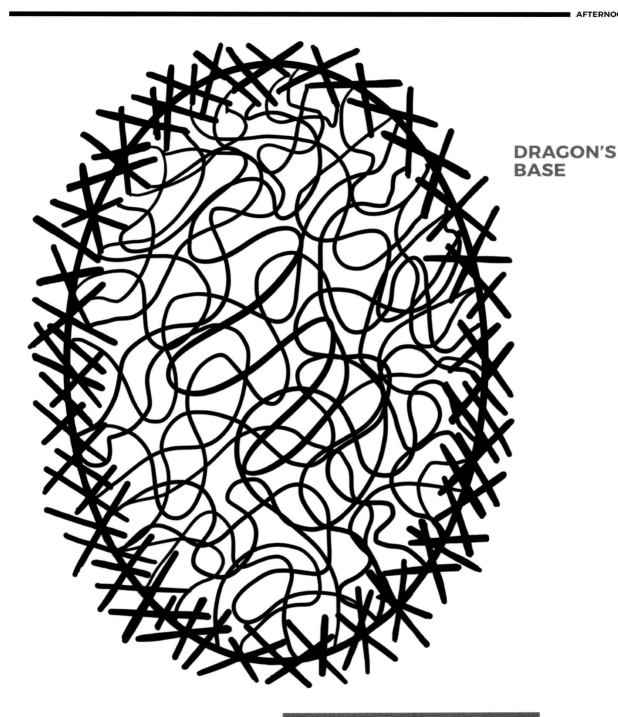

DRAGON'S BASE

MAKE A BASE FOR YOUR DRAGON

Drawing circular and oval shapes can be quite difficult so here we use straight lines to form the outline. Once you have completed the outline use the filigree infill technique.

JOINING THE DRAGON BODY TO THE BASE

Decide where to position your dragon. Initially we attach the legs to the base with the spot welding and reinforcing techniques. Twist and bend the tail and fix the end to the base in the same way. ◻

EXTENSION IDEAS

Make your dragon in different colours. Add flames to the mouth and maybe spikes or horns to the head or body. Lighting brings a 3D pen sculpture to life. I usually use a variety of small, battery-powered, wired string LED lights. Add these to the back of the base, and then add a hanging ring and turn your 3D pen Dragon into a wall art light. You could probably use lights controlled by a Raspberry Pi!

Above ⬉
The base gives a solid foundation to your build

TUTORIAL ━━━━━━━━━━

Control a K40 laser cutter over a network

The easy way to send jobs to your cutter

Steve Pelland

🐦 @StevePelland

Steve works in IT by day and can be found in his garage shop at night. When not in his shop he can be found at the local hockey rink watching his son play. He enjoys working with the Scouts teaching STEM badges.

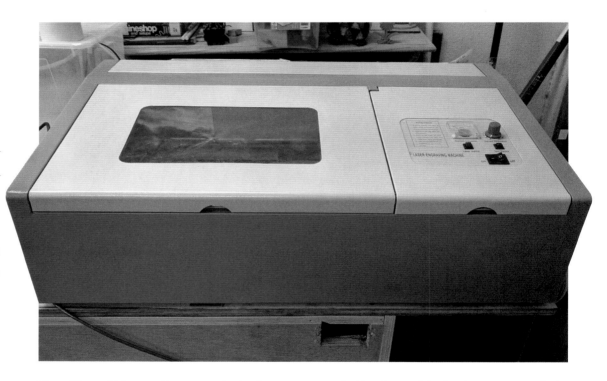

Above ▨
K40 lasers can be an affordable way to start cutting

K40 laser cutters are a staple of workshops around the world, as they're one of the most affordable routes into computer-controlled manufacturing. However, by default, they're very bare-bones, and it takes a bit of work to get a setup you're comfortable with. One problem with the off-the-self setup is that you have to leave your computer plugged in via USB to run your job. Let's take a look at one way of removing this dependency, allowing you to run jobs without your main machine plugged in. This is particularly useful if you've got a desk in your workshop away from where your laser cutter is set up.

Setting your laser cutter up like this will let you control it from anywhere. However, we can't recommend this. Laser cutters essentially run via controlled burning, and the line between controlled and uncontrolled burning is slim. Even when running a job you've done before, it doesn't take much to set fire to your machine. This setup means you can keep your computer in a more convenient part of your workshop, so you can keep an eye on your machine. And, of course, don't forget to keep a fire extinguisher on hand in case something does go horribly wrong.

We tested this using a Raspberry Pi 3B+ with Raspbian Stretch June 2018 installed. In theory, any RPi should work, but we'd recommend either a 3B+ or a 3A+ as you'll get better performance. You'll need this setup before starting the tutorial.

It's easiest to set up the RPi with a mouse, keyboard, and monitor attached, but we'll remove these once everything is running.

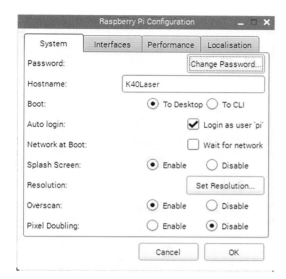

Once your machine has booted, run the Raspberry Pi Configuration application from the main menu – click on the raspberry icon in the upper left-hand corner of the screen – then click on Preferences and then Raspberry Pi Configuration.

Under the System tab, you can set the host name, change password, etc.

The important part is in the Interfaces tab, where you can enable the camera, SSH, and VNC. For this tutorial we only need to enable SSH.

It makes it easier to keep track of your Raspberry Pi if it has the same IP address every time it boots. Normally these are assigned by your router every time your turn your machine on. On some networks, you can set the static IP directly on your RPi. To do

this, right-click the WiFi icon in the upper right-hand corner and configure your chosen interface. However, on some networks you'll have to make this change in your router's configuration – this is a little different for each router, so check your manual if you're not sure.

Now we can shut down the RPi, disconnect the keyboard, mouse, and monitor, and move it next to the laser cutter.

Upon booting, use your SSH client of choice. If you don't have an SSH client of choice, PuTTY or MobaXterm works fine for Windows (we like MobaXterm because of the tabbed connections).

Our setup in MobaXterm is as follows:
- The remote host is the static IP address that you set up previously
- The default username and password are Username: pi and Password: raspberry
- Enable X11 Forwarding
- Click OK

Double-click the session on the left-hand side if you are using MobaXterm or connect with whatever client you are using. Enter your password when prompted, and you should be dropped into a command prompt.

You'll also need some software to display the images on your local machine. This is known as an X server. If you're using Windows, Xming (**sourceforge.net/projects/xming**) works well, and if you're on a Mac, XQuartz (**xquartz.org**) is a good option. →

Left ◈
We named our RPi K40Laser so it's easy to identify on the network

Below ◲
Setting a static IP helps you keep track of your machine across reboots

Below ◈
You can turn on or off a variety of low-level protocols on your Raspberry Pi

YOU'LL NEED
◈ **Raspberry Pi**

  K40 laser cutter

◈ **A desktop computer**

◈ **A display you can temporarily plug your RPi into for the OS installation and configuration**

◈ **USB cable** (should have come with your K40)

Control a K40 laser cutter over a network

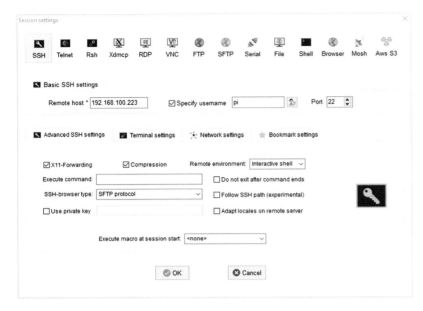

That's the basics set up. We can now install the software we need to control the RPi. SSH into your machines and run the following:

```
sudo apt-get update -y
sudo apt-get upgrade -y
sudo apt-get install -y libxml2-dev libxslt-dev
sudo apt-get install nano
sudo apt-get -y install inkscape
```

You'll now need to turn on your K40.

We will need to find out the USB device ID of the K40: we can use the `lsusb` command. When executed, we received the following output (but yours will be slightly different):

```
Bus 001 Device 011: ID 1a86:5512 QinHeng
Electronics CH341 in EPP/MEM/I2C mode, EPP/I2C
adapter
```

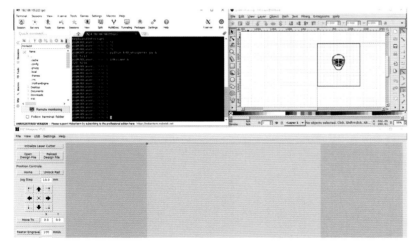

Copy this to a text editor for future use. In this example '1a86' is the VENDOR ID and '5512' the PRODUCT ID.

Create a udev control file for your laser cutter. You'll need to do this using a text editor running as root, which you can launch with:

```
sudo nano /etc/udev/rules.d/97-ctc-lasercutter.
rules
```

Put the following text into the file and replace [VENDOR ID] and [PRODUCT ID] with the information you obtained from lsusb:

```
SUBSYSTEM=="usb", ATTRS{idVendor}=="[VENDOR
ID]", ATTRS{idProduct}=="[PRODUCT ID]",
ENV{DEVTYPE}=="usb_device", MODE="0664",
GROUP="lasercutter"
```

Press **CTRL+X** to save and exit. You'll need to reboot your RPi to make these changes take effect.

Once it's back up and running, you'll need to SSH in again and download the K40 Whisperer source code, for example **K40_Whisperer-0.XX_src.zip**.

Use the `wget` command to download the file:

```
wget http://www.scorchworks.com/K40whisperer/
K40_Whisperer-0.17_src.zip
```

Version 0.17 was current at the time of writing, but any later version should work.

Once downloaded, let's unzip to a directory.

```
mkdir /home/pi/K40
unzip K40_Whisperer-0.17_src.zip -d /home/pi/K40
```

Go to the **K40 Whisperer** source directory, and install the required packages:

```
cd /home/pi/K40/
pip install -r requirements.txt
```

Everything's now installed and ready to run, but we just need to do a bit of jiggery-pokery to get the windows to appear in the right place (on our main machine, not on the monitor that's no-longer attached to the RPi). This means using `xauth` to point to the right place. We've written a script to do it all for you. Save this in a file called **sudox** in your home directory:

```
su - pi  -c 'xauth list' |\
    grep 'echo $DISPLAY |\
```

```
        cut -d ':' -f 2 |\
        cut -d '.' -f 1 |\
        sed -e s/^/:/' |\
 xargs -n 3 xauth add
```

Now make this file executable with:

```
chmod a+x /home/pi/sudox
```

One more reboot should make sure everything is set up and ready to run.

Once rebooted and logged in via SSH, you can change directories into the **K40** directory and launch K40 Whisperer.

```
./sudox
cd K40/
python k40_whisperer.py &
inkscape &
```

(The **&** symbol tells Linux to run the program in the background, allowing you to issue more commands with another prompt.)

Now that we are up and running, we need one more thing. How do we send our jobs to the RPi? You can just open a browser from the RPi, download items, work in Inkscape, and then run in K40 Whisperer. What if you designed something on your workstation and just need to send it to K40 Whisperer? Easy. Let's just set up a share on the RPi, map the drive in Windows, and then you can drag and drop or even just save files directly to the share. Then you have the freedom of choice – work from the RPi or work in conjunction with your workstation.

So, let's get Samba installed. Samba will allow us to share our work directory easier with Windows clients.

```
sudo apt-get -y install samba samba-common-bin
```

Let's create the directory:

```
mkdir ~/K40Projects
```

Now we need to edit the Samba configuration file:

```
sudo nano /etc/samba/smb.conf
```

And add the following to the bottom of the file:

```
[PiShare]
comment=Raspberry Pi Share
path=/home/pi/K40Projects
browseable=Yes
```

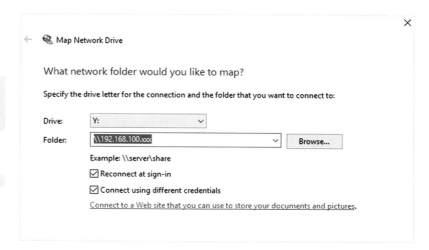

Above ◈
Link your two computers together with file sharing

```
writeable=Yes
only guest=no
create mask=0777
directory mask=0777
public=no
```

Now you can set the password for file sharing. Enter the following in a terminal:

```
sudo smbpasswd -a [username]
```

ENTER THE PASSWORD TWICE

Now you can connect to this from your Windows machine. Go to Windows Explorer, right-click on My Computer and select Map Network Drive.

For the folder, use **\\[ip address of pi]** and then click on Finish.

You will then see a password prompt. Enter your password. And your drive should be mapped.

That's everything set up and ready to you. You can now use your laser cutter from other machines – the only thing you have to decide is what to cut next. ▢

MORE INFORMATION

We learned lots from the online community of Raspberry Pi and laser cutters. You can find out more about the process we went through here at the following resources:

- Setting up K40 Whisperer and Inkscape on the RPi
  raspi.tv/2018/run-a-k40-laser

- Forwarding applications utilising sudo through an SSH tunnel
  joelinoff.com/blog/?p=729

- Creating shares on Raspberry Pi
  raspberrypihq.com/how-to-share

TUTORIAL ━━━

Playing (tunes) with your food

Make a touch-sensitive cake that plays music

Ben Everard

🐦 @ben_everard

Ben loves cooking stuff – his fridge is packed with a thousand highly experimental dishes which only he will end up eating...

Food, music, and electronics are three of our favourite things, so it only makes sense to attempt to combine them all into one delicious, melodious circuit – surely? We decided to try to make a touch-sensitive cake, on which you could actually play different notes by touching different parts of it, like a kind of tasty edible piano.

You can do this with more or less any controller that can take touch input (regular GPIO pins won't work; they have to be touch-sensitive). We tried this with an Adafruit Circuit Playground Express and a Bare Conductive Touch Board, and both worked without problems. But before we get to the controls, let's take a look at the tasty part of the build – the cake.

The main challenge of this build is that we want a single cake with multiple parts on it which respond to touch. Simply putting the electrodes in the cake doesn't work, because cakes aren't very conductive. To make this work, we need a way of getting a conductive electrode on the surface of the cake. We could just put an electrode on there – there are plenty of food-safe electrodes, such as aluminium foil – however, we want the entire cake to be edible, and this meant creating an edible electrode.

There are some conductive doughs that use salt and/or lemon juice; however, they stretch the definition of edible – they might not outright kill you, but they're going to taste horrible and, if eaten in quantity, might make you throw up. Not exactly

qualities we look for in a cake. That left one option: edible metal foil. Pure forms of silver and gold are both edible and – when hammered into fine foil – malleable enough for cake decoration. We believe both gold and silver foil should work, but we used silver leaf because we're cheap, and it's a better conductor anyway.

While both gold and silver are expensive metals, they can be beaten incredibly thin – often around 0.1 micron. To put this in perspective, you'd have to stack 750 of these sheets on top of each other to get the thickness of a human hair.

> **Gold and silver leaf comes in a few forms** – most commonly, loose leaves and transfer sheets

Gold and silver leaf comes in a few forms – most commonly, loose leaves and transfer sheets. In principle, both should work. We opted for transfer sheets, which are leaf metal on tissue paper. This allowed us to cut the leaf into strips the right width for our keys before applying.

GILDING THE LILY

We bought a pack of 25 silver leaf transfer sheets for £9.85, and we used five in this cake, so the total silver cost was a measly £2.

Do be careful when buying edible leaf though, because there are many types of non-edible imitation and genuine, but less pure, silver and gold leaf around. Typically, these contain non-edible metals (such as copper), so always buy from a reputable source and ensure that it's marked as edible.

You can use a wide variety of cakes, but they do need to have firm – and ideally fondant – icing. Our

first attempt was on a cake with butter icing, but the leaf is so delicate that the slightest movement rips it apart and this softer icing couldn't support it well enough to create a continuous conductor. Fondant is firm enough to hold the leaf in place and just sticky enough to adhere to the leaf.

It's hard, but not impossible, to get the leaf to adhere around shapes and strange contours on the cake. The flatter the edges of the cake, the easier it will be to coat, so it's best to start with a simple design. We cheated and bought a supermarket cake – this isn't a baking tutorial after all. We'll leave it up to you whether or not to start from scratch.

With all that out of the way, let's get on to making the keyboard cake. First, you're going to need some electrical contacts for the cake to sit on. This is what you'll wire up to your touch-sensitive board. You need to start with a non-conductive base, such as a plate (note that many cakes come on foil bases that are slightly conductive). You'll need to move your cake off this onto something non-conductive before starting. →

Left
Silver leaf transfer sheets enable you to shape the silver before applying it to the cake

Above
The cake-to-plate transition is the trickiest part of the wiring, but did work after some practice

CAPACITIVE TOUCH

Capacitive touch essentially works by sensing the capacitance of a particular output. When you touch it, your finger affects the capacitance and this is detected and processed as a 'touch' that you can process in your code. This is different from general input/output pins that need a voltage applied in order to trigger an action. On some devices, I/O pins are both normal pins and capacitive touch sensors. You may need to set them up differently in your code. You'll often find that touch-sensitive inputs are larger conductive pads designed for crocodile clips rather than 'pin' headers. However, this isn't a universal thing.

YOU'LL NEED

- ◈ **Cake**
- ◈ **Fondant icing**
- ◈ **Silver leaf**
- ◈ **Aluminium foil**
- ◈ **Touch-sensitive board**

TUTORIAL ━━━━━━━━━━━━━━━━━━━━━━━━━━━━━━━━━━

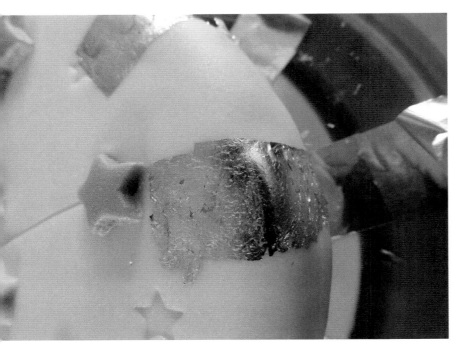

silver leaf. The number and placement of these pads is up to you. Take a moment to look at the cake and see how you want the 'keys' laid out, and then work backwards from this to where you want the pads. You'll need at least 3 cm between the pads to avoid any cross-talk. If you want to play *Happy Birthday*, for example, you'll need at least seven notes, as it doesn't use sharps or flats.

Once you've got these in place, pop your cake on top and it's time to get gilding.

MUSICAL DECORATIONS
You start with silver leaf on the transfer paper and you want to end up with it on the icing. The first part of this is making the icing a little stickier. Brush it lightly with water. You don't want it to be wet, but just damp enough to be tacky. Then cut the transfer paper up into strips about 1–2 cm wide and 2–3 cm long (this will depend on the pattern you want on the cake, but this width worked well for us). It's much easier to shape the leaf while it's still on the paper than trying to apply a whole leaf but with only some sticking. You should then be able to lay your transfer paper down on the damp icing, rub the back of the paper (scratching can help too), and peel it back, leaving the silver leaf on the icing. You'll often find that you get some leaf that's come off the transfer paper but hasn't properly stuck to the icing. It's handy to have a clean brush to push it onto the icing with (but if you don't have one, a clean finger will usually do the job).

On this non-conductive base, you need some 'pads' that form the interface between your Touch Board and the silver leaf on the cake. Aluminium foil is the obvious choice, so start by cutting out sections of foil about 2 cm by 5 cm, and taping them in place on this base. They'll need tape in the middle so that the cake is sitting on bare foil, and there's also bare foil to attach a crocodile clip to on the opposite side of the tape to the cake – this is so the crocodile clip won't move the foil and break the connection with the

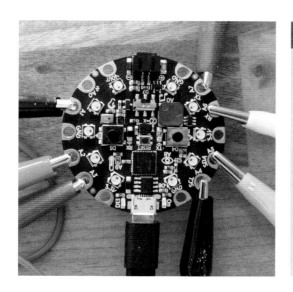

Left ◈
The internal speaker on the CPX isn't very powerful, but you could attach it to an external speaker between A0 and ground

THE **EASY WAY**

There are many, many ways this project could be done far more easily. A large amount of the problem is the fact that we're converting a single object into a completely edible multi-touch interface. If you're willing to drop either the single object or the entirely edible, then the whole thing becomes much simpler.

Fruit is conductive enough for capacitive touch and you can wire up different fruit to a capacitive touch interface by simply sticking an electrode in them. Get seven fruits and seven electrodes and you can create the same musical interface as our cake, with no gilding required. Alternatively, if you're willing to include non-edible items on your cake, then everything becomes simple. Perhaps some foil-covered chocolate coins could be connected to the touch inputs, or simply aluminium foil cut into an interesting pattern.

That, at least, is the theory. It's a bit tricky in practice because the leaf is so delicate, and can at times seem to refuse to stick to the icing for no discernible reason. You'll almost certainly find that you don't get a continuous trace from this, so you'll have to go back over some areas that you've applied leaf to and double or triple-up the leaf. This is to be expected, and using this, you can make traces longer than a single strip of transfer paper.

The hardest part of the endeavour is getting a contact between the silver leaf and the aluminium foil 'pads'. The most reliable way we found of doing this was to slightly crinkle the transfer paper before applying this section: that helps it almost fall off the transfer paper once it's pressed in, and then a gentle brushing (or prodding) will get it in place. Bear in mind that this connection will be very fragile, so if you plan on moving the cake more than a short walk, it might be best to do this after you've moved it.

TOUCHY FEELY

The last part is just to wire the cake up to the touch sensor. This is just a case of clipping an alligator lead to the aluminium foil and attaching the other end to the touch pad on your chosen board.

When it comes to software, there's absolutely no point in writing your own unless you want something specific. There's almost certainly some pre-existing music setup for your board. If you've gone with the CPX, the Adafruit Lime Piano code works and can be found at **hsmag.cc/HMkREe**, while if you've gone with the Bare Conductive Touch Board, you should find a MIDI example in your Arduino Sketchbook if you follow the installer here: **hsmag.cc/RsFoyH**.

If you've used MIDI code you're also going to need MIDI software on your computer to turn these signals into sound. We used ArcTrax, which is available in the Windows store, but any software that can take MIDI input should work. Just make sure your Bare Conductive Touch Board is connected, and select this board as 'MIDI in', and then you should start to hear melodic sounds when you touch your edible, electric cake. You should find similar software for other touch-sensitive boards too.

That's all there is to it, so what are you waiting for? Life's a bit nicer with cake, and even better with electrical cake, and the music is waiting for you – so don your apron and get baking. ☐

Below ◱
All together now…
Ha-ppy birth-day tooooo yooouuuu

REGULAR ▬▬▬▬▬▬▬▬▬▬▬▬▬▬▬

Rubik's Cube solver

By **André Angelucci**   ⊘ hsmag.cc/mJpcxn

" **J** **ust like computing and robotics, the Magic Cube is a puzzle that involves reasoning, geometry, calculations, and dedication.** That motivated me to develop a robot, as the work of completing my degree in Computer Engineering, capable of solving it.

"Basically, its operation involves software responsible for calculations, and a robot that physically solves the puzzle.

First, the software captures photos of each face of the cube and identifies the colours, through IBM's artificial intelligence API, Watson. With the recognised colours, a calculation is made that results in a sequence of approximately 15 movements that solves the puzzle, using Herbert Kociemba's two-phase algorithm. The solution is sent to the Rubik's Bot, the name of which is a tribute to the creator of the puzzle. An ESP32 microcontroller receives the message, and distributes the motions between six NEMA 17 stepper motors which, through iron rods, physically perform the rotations on the faces of the cube to complete the solution in about five seconds." ❑

Left ◈
What next for our
laser-cut overlords?

Way Home Meter

Use an ESP8266 and some NeoPixels to let loved
ones know when you'll arrive back home

Brian Lough

🐦 @witnessmenow

Brian is a maker from
Ireland who primarily
creates projects
and libraries for ESP
microcontrollers.
Check out his stuff on
his YouTube channel
and **blough.ie**

" **L** et me know what time you'll be
home"– it's a common refrain in
homes across the country. We try
to give good answers, but it's hard
to know how traffic will affect us on
the way. Rather than rely on guess-
work, let's try to build something to let our families
know when we'll make it back.

In this project we'll build a device that will give
up-to-date home arrival times, based on the live
traffic conditions. To make the device more useful
for when it's not being used for that purpose, it
works as a clock that automatically fetches its time
from the internet and also automatically adjusts for
daylight savings.

We've built this using an ESP8266, which is a
surprisingly powerful microcontroller with built-in
WiFi and can be programmed using the Arduino IDE.

The device makes use of a few different, free
internet services:

- **Telegram:** an instant messaging service that
  allows for the creation of bots that users
  can interact with. It is a really good way of
  communicating with your ESP8266 or ESP32
  projects from anywhere in the world, for free.

- **Google Maps API:** can be used to get travel time
  and traffic information between two places.

- **NTP servers:** Network Time Protocol, a way for
  network-connected devices to get the time. This
  saves the needs for a real-time clock, and also
  doesn't require the time to be set.

To use it, the person who is coming home uses
Telegram on their phone to share their live location
to a Telegram Bot that is running on the Way Home
Meter. This will update the Way Home Meter
with the person's GPS coordinates every 20 or
30 seconds.

The Way Home Meter takes these coordinates
and sends a request to the Google Maps API to
get the live travel time and distance between the
person's location and home.

The Way Home Meter will then add the travel
time onto the current time and display the estimated
arrival time of the person and updates the dial and
NeoPixels to represent what percentage of the
journey (distance wise) has been completed.

CODE IT UP

The code for this project is available on GitHub.
Go to the following URL, **hsmag.cc/ybAcHB**,
and click the Clone or Download button on the
right side of the page, and then Download Zip.

Left ◈
Screw terminals are
useful for projects
where components
are separated from
the PCB

YOU'LL NEED

◈ **An ESP8266 Wemos D1 mini microcontroller** (or equivalent e.g. Adafruit Feather Huzzah or NodeMCU etc.)

◈ **4-in-1 Max7219 dot matrix display**

◈ **A small servo** (sg90)

◈ **A 3D-printed dial for the servo** (optional, could be made from anything!)

◈ **11 × Through-hole NeoPixel** (I used PL9823 LEDs)

◈ **220 pF capacitor**

◈ **Passive buzzer**

◈ **1 kΩ resistor**

◈ **NPN transistor**

◈ **Protoboard** (I used a prototype PCB I designed, but the project can easily be built with standard protoboard)

◈ **Screw terminals** (optional)

◈ **IKEA RIBBA frame**

◈ **A3 piece of 3 mm foam board**

◈ **Hot glue gun**

◈ **4 mm wood drill bit**

◈ **Sharp knife**

◈ **A metal ruler**

◈ **A compass and a protractor**

◈ **Micro USB phone charger** (for powering the project)

Extract the zip file. Inside the extracted folder, open up the **WayHomeMeter** folder and open the **WayHomeMeter.ino** file.

This sketch requires some additional Arduino libraries to be installed; start by opening the Arduino Library Manager by going to Sketch > Include Library > Manage Libraries.

You will need to add the following libraries:

- **Universal Arduino Telegram Bot** by Brian Lough – for creating a Telegram bot on the ESP8266.

- **Google Maps API** by Brian Lough – for getting the live traffic data.

- **Arduino JSON** by Benoît Blanchon – used by the libraries to parse the responses. Note: There is a breaking change in V6 of this library that will cause it not to work with the Telegram and Google Maps library, so use the drop-down on the left of the window to change the version to V5.13.2.

- **MD_MAX72XX** by majicDesigns – for communicating with the dot matrix display.

- **MD_Parola** by majicDesigns – handles animations on the dot matrix display.

- **Adafruit NeoPixel** by Adafruit – for controlling the NeoPixels.

- **NTPClient** by Fabrice Weinberg – for getting the time from the internet.

- **Timezone** by Jack Christensen – for automatically switching the time for daylight savings. →

PROGRAMMING THE **ESP8266**

The standard Arduino IDE isn't set up to program the ESP8266, so before we can program the board, we need to set this up (you can skip this bit if you've already used the IDE with an ESP8266).

First, let's get the raw IDE. You can download this from the Arduino website and install it as you would any other software: **hsmag.cc/TAfEJp**.

Next, you will need to set up the IDE so it knows how to communicate with an ESP8266. Open the Arduino IDE, go to File > Preferences, and paste the following URL into the Additional Boards Manager URLs, then click OK: **http://arduino.esp8266.com/versions/2.4.2/package_esp8266com_index.json**

Back on the main screen of the Arduino IDE, go to Tools > Board > Boards Manager. When this screen opens, search for 'ESP8266' and install it; this may take a few minutes depending on your internet connection.

After setting up a new board it is recommended to get the simple example blink sketch before trying anything more complicated; this can save a huge amount of headache down the line! You can find this in File > Examples > 01. Basics > Blink.

Upload this to your ESP8266 and you should see an LED blink on and off. If you get an error or don't get a blinking light, make sure you've got everything installed correctly and the ESP8266 is properly connected.

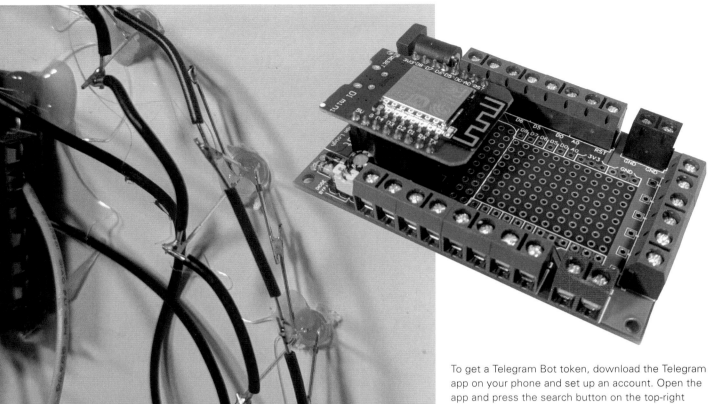

Above right ⬈
This is a custom PCB that breaks out all the pins of the D1 Mini to screw terminals, but it can be easily recreated with a standard protoboard

Above ◈
The address pins of the LEDs should be able to reach each other without the need for extra wire. The power pins need to be joined by wire

After installing these libraries, you should click the 'verify' button (shaped like a tick) on the WayHomeMeter sketch to make sure that everything compiles fine.

SOME CONFIGURATION REQUIRED

You will need to make some configurations to get this sketch to work for you, but you will first need to get:

- Telegram Bot token
- Google Maps API token
- GPS coordinates of your home

GOOGLE BILLING

Google Maps gives a free monthly allowance of credit – equivalent to 20,000 requests. That is just under what's required to send a request every two minutes in a month (about 22,000). This device only makes the request every two minutes that it is actively monitoring someone's home journey, so should stay under the limit if used occasionally. It's possible this limit will change in the future. How often it checks can be configured in the sketch by changing **delayBetweenGoogleMapsChecks**.

To get a Telegram Bot token, download the Telegram app on your phone and set up an account. Open the app and press the search button on the top-right of the screen. Search for 'botfather'. Type /newbot and follow the on-screen instructions. The botfather will provide you a link to the bot and an access token. The link is for the chat where people will share their location; the access token is used in the sketch to authenticate your ESP8266 as the bot you just created.

Next, you will need to get a Google Maps API key. Start by going to the following URL: **hsmag.cc/mPqFqh.**

Check the Routes option and click Continue. You will then be asked to create a project; you can give this any name. You will need to add a billing account, but this device will comfortably operate on the free allowance given by Google. You will then get an API token that can be used in the sketch.

And finally you will need to get your home's GPS location. A simple of way of doing this is using Google Maps. Using a web browser (not the app), navigate to your house on Google Maps and right-click and click 'Directions from here'. This will modify the URL, which will now contain the coordinates of your home; copy and paste these from the URL e.g. 51.5546466,-0.2794867.

You now have everything you need to configure the WayHomeMeter. Open up the WayHomeMeter sketch and click on the config.h tab. First thing you will need to enter is your WiFi details so that the ESP8266 is able to connect to your WiFi.

NEOPIXELS WITH A 3.3 V DEVICE

You can often have issues using NeoPixel LEDs with a 3.3 V logic level device such as an ESP8266 or Raspberry Pi. You can get around this issue by using a logic level shifter to convert the 3.3 V to 5 V for the Data In connection for the first LED. However, we've found that it works fine with just a small capacitor between Data In on the first LED and Ground (as seen in this project).

Next, you will need to add your Telegram Bot token, your Google Maps API key, and your home location. Finally, if you are not in the UK or Ireland, you will more than likely need to change your time zone. Uncomment the appropriate time zone and comment out the UK and Ireland time zone.

WHAT MAKES IT TICK

You'll need to wire everything together as shown in **Figure 1** (overleaf). The LEDs are addressable RGB LEDs, so they only require a single GPIO pin of your microcontroller and you can set the colour of each LED individually. The input of the first LED (the one over on the left when looking at it from the front) will be connected to the Wemos, and its output will be connected to the input of the second LED. For all subsequent LEDs, the input of the next

> Once you are happy everything is working correctly, **secure the components in place with some hot glue**

LED is connected to the output of the previous one. The output of the final LED will not be connected to anything. Soldering these LEDs should be left until the final assembly stage.

CREATING THE MOUNT

Take the back panel off the picture frame and use it to trace a square onto your foam board. Using a sharp blade, cut out the square.

Next, you'll need to separate the panels from the display, as you'll be placing the PCB on the back side of the foam and the panels on the front; this will hold the display in place and hide the cuts from view.

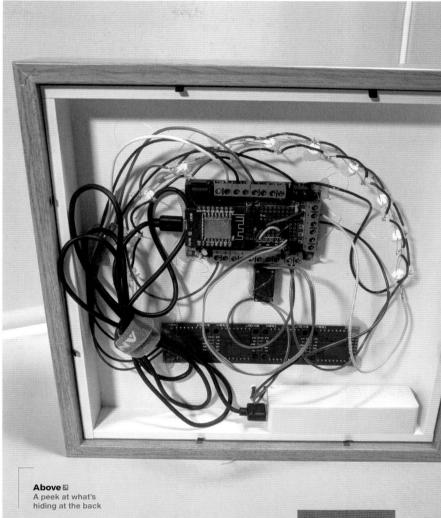

Above
A peek at what's hiding at the back

Carefully remove each of the dot matrix panels from the dot matrix display. There are markings on the side of each of the dot matrix panels; make a note of what direction they are facing in comparison to the PCB to ensure they are put back in the correct orientation.

If your PCB has header pins attached, desolder and remove them. Replacing these with wire will make the PCB fit flush to the foam board.

Measure the rectangle created by the pins and mark out that shape where you want to place it on the foam board. The objective is to cut out a shape that the pins of the PCB will fit through, but the PCB itself will be too big for.

The LEDs for this project are all on an arc around the centre point of the servo. Mark where you want the centre point of the servo arm to be and, using your compass, draw a semicircle lightly for where you want the LEDs to be. →

SHORT ON TIME?

If you are short on time, or just interested in quickly trying this project out, strip it back to be just the dot matrix display, the Wemos D1 Mini, and use DuPont cables to connect them together. The key piece of functionality, displaying the expected arrival time, uses only the display.

QUICK TIP

It's a good idea to practise cutting out the shapes and making the LED holes using scrap pieces of the foam board!

Place your protractor on the centre point and mark every 18 degrees. Then, using a ruler, line up the centre point and these new marks; where this line intersects with the semicircle is where each LED should be placed. Starting on the side that you want to be the front, use the 4 mm drill bit by hand (no power drill needed) to create a hole for each of the LEDs where you have marked.

Measure the dimensions of your servo and mark it on around the centre point. Remember that the part of the servo that rotates should be the centre point, so offset the servo shape to suit. Cut the shape out of the foam board and place the servo through from the front.

Finally, you will need to place the buzzer module. You can simply place the pins of the module into the foam to mark where the holes should be and, using a piece of wire, pierce the two holes so they go through the foam board.

ON THE FINAL STRETCH

Place all the LEDs into the foam board from the back. Bend the input pin of each LED back towards the previous LED, and bend the output pin of each

GOING FURTHER

You can take this project and make it differently, depending on your requirements. Here are a few suggestions:

● Add the ability to send the device a location and time, and have it calculate when you need to leave your home to make it on time.

● Add support for multiple people. Currently the device will request the travel time for the person who last sent a coordinate and display the correct name and information for them, but this could be improved to handle multiple people.

● Configurable alarms. Get notified when a person is X number of minutes away. Useful for starting dinner!

towards the input pin of the next LED, and solder them together. Slightly bend all the Ground pins of the LEDs towards the centre of the circle and all VCC pins away from the centre of the circle. Solder wire between all the Ground and VCC pins.

Place the dot matrix PCB in the cut-out, and put all the panels back in place. Pay careful attention

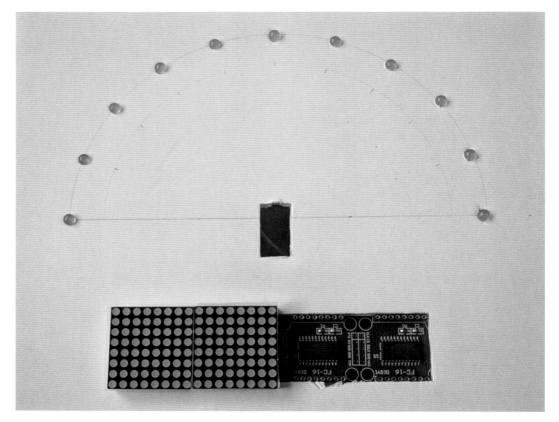

Right ◈
Foam board all cut and LEDs placed. The foam board will be clamped between the display module and the PCB

A POTENTIAL WEASLEY CLOCK?

A lot of people who saw early versions of this project mentioned that it reminded them of the Weasley Clock from *Harry Potter*, a clock that showed the current location of each of the members of the Weasley family. This Telegram-based solution could be used for a project like that, but it does require each of the users to actively enable the location sharing. A more passive solution might be better.

QUICK TIP

Always be generous with the lengths of wire you use, especially in a project where space is not an issue. If you need to make adjustments, it's easier to shorten them than lengthen them.

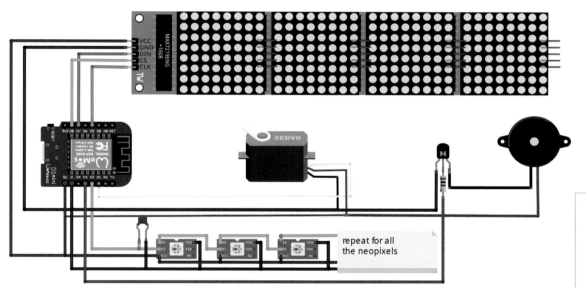

repeat for all the neopixels

Figure 1 ◈
The wiring diagram for our Way Home Meter

Below ◈
The name displayed comes from the user's Telegram name

to the orientation of the panels, as it is very difficult to remove these again without damaging the foam board.

You then want to thread the wire of the servo module through the hole for the servo, and then insert the servo. Glue the dial hand onto one of the connectors that comes with the servo, and attach it to the servo when dry. The 3D design used in this project can be downloaded from here: **hsmag.cc/iqOPiP**. However, you can use anything you want (and a model car could be substituted if you don't have access to a 3D printer).

Finally, solder wire to each pin of the buzzer module and push it through the front of the foam board.

Connect all the modules to the Wemos on the protoboard and test everything out. Once you are happy that everything is working correctly, secure the components in place with some hot glue. You are now ready to have super-accurate home arrival times! □

TUTORIAL ▬▬▬▬▬▬▬▬▬▬

Monkey skull treasure bowl

Anyone who is daring enough to reach for treasure from this bowl will be in for a big surprise

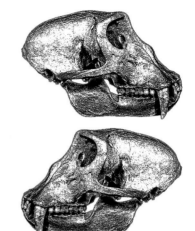

Andy Clark

🐦 workshopshed

For the last 10 years Andy has been making and repairing in a shed at the bottom of the garden. You can see more of his exploits at **workshopshed.com**

The Kuriologist

🐦 @strangecurios

A self-taught cognitively disabled 'outsider' artist. To overcome housebound boredom, creates mixed media sculptures inspired by 'cabinets of curiosities' and dead stuff. **kuriology.com**

We teamed up with artist 'The Kuriologist' to make this solder-free project, inspired by the *Indiana Jones* films. Despite having a hollow head, this papier-mâché skull bowl actually has its own brain, a tiny Adafruit Gemma M0 board. We wired that up to some bejewelled LEDs using crimp connectors. The rim of the bowl hides the wire for a capacitive sensor which triggers our spooky skull.

MAKING THE SKULL

The base of the skull is made from a used tape roll. On top of that, place a small box which will be used to house the electronics. The opening part of the box should be positioned to the back of the skull, use masking tape to secure. A second roll taped on top gives the skull some height and forms the bowl. Scrunch the newspaper into packets of a 'sausage roll' shape.

You can then tape the packets to the rolls to form the rough outline of the skull. The jaw can be cut from thick card and taped to the front of the skull. Cut nasal and eye sockets from card and add with more tape. Keep poking, squeezing, adding more masking tape, and building up areas until it looks like a skull.

Mix up some PVA glue, three parts glue to one part water. The water will help the glue soak into the paper and improve drying times. Rip up white paper

Right ◈
Monkey brains anyone?

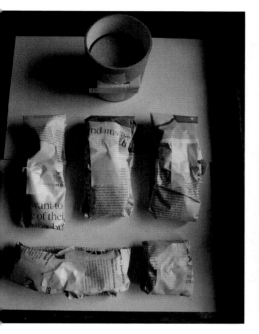

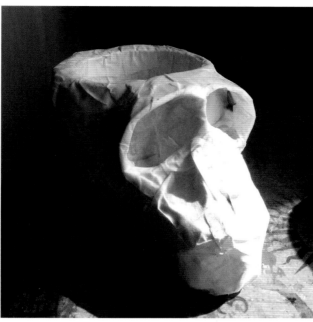

Far left ◈
A solid base for the
skull and padding
to shape

Left ◈
Skull shape formed
from tape and card

QUICK TIP

The 38 mm masking
tape is best, as it is
not too wide and not
too narrow.

and dip into the glue mix before applying all over the skull. Ensure it is fully covered, and apply two or three layers for strength. It's a good idea to leave the skull 30 minutes to one hour between each layer to dry. Then leave overnight for it to harden.

Once it's dry, we need to decorate the skull. This can be done with a final skin of paper using a printed image of a skull, a technique called découpage. An example skull image is available on Wikimedia Commons: **hsmag.cc/eBBMCv**. Apply this in the same way as the papier-mâché layers, by ripping the picture into smaller pieces and sticking them onto the skull. Use plain black paper for the eye sockets and red paper for the bowl. An alternative would be to paint or colour these with a permanent marker.

To give the skull a protective skin, it is coated with shellac. The 'button' or 'garnet' colours work best. It dries quickly and gives an antique appearance. One coat should be sufficient. If you use a dedicated brush for shellac, you can give it a quick rinse in thinners before storage, and then a soak before the next use.

GLOWING GEMSTONES

Select some acrylic gemstones of a suitable size to fit into the eye sockets. The stick-on or stitch-on type, that have a flat side, should be easiest to use. The flat side is painted silver and helps reflect the light. So that the light from the LED will shine through, lightly sand the silver paint.

A reflector cone is formed from kitchen foil. To ensure there is some overlap to join the cone, we needed to draw our template slightly oversized. Draw a 15 mm diameter circle, and a second circle of 40 mm around that. Cut out the large circle and slit the disk to the centre. Now cut out the inner circle.

Form the foil into a cone and wrap around a 10 mm LED. Tape this into place. If your gemstones are not round, you may need to make some adjustments to the shape of the cone. You can do this by folding over the edges, pinching and folding, or by cutting off the excess. Fill the cone carefully with hot glue and stick the gemstone to the top and leave to cool. Repeat with the other LED and gemstone.

WIRES AND CRIMPS

You will need two lengths of different coloured wire for each LED, which are long enough to pass through the skull and out of the back. This will mean you can connect the Gemma before sliding it back into the box. Crimp one wire to each leg of the →

Above ◈
The reflector helps spread the light across the gemstone

YOU'LL NEED

◈ **Newspaper**

◈ **Small cardboard box**

◈ **Paper and card**

◈ **Used tape rolls**

◈ **38 mm masking tape**

◈ **PVA glue**

◈ **French shellac polish**

◈ **2 × Jewels**

◈ **2 × 10 mm red LED**

◈ **2 × 220 Ω resistor**

◈ **Kitchen foil**

◈ **Hot glue**

◈ **Adafruit Gemma M0**

◈ **Wire**

◈ **Approx 30 cm of screened wire**

◈ **Crimp connectors**

◈ **3 × AAA battery box with JST connector**

◈ **2.5M nuts, bolts, washers**

TUTORIAL ━━━━━━━━━━━━━━━━━

LED. Make a note of which wire connects to the anode of the LED.

Use a bradawl or skewer to make a hole in the back of the eye sockets and into the cardboard box. Pass the wires through the holes and out of the box. Crimp a resistor to one of the LED wires.

YOU'VE GOT THE TOUCH

The sensor for the touch is formed from the screening from the screened wire. Remove the outer plastic cover of the screened wire and slide the braided screen off the wires. Flatten the braid and form into a loop to fit around the rim of the bowl. Twist the ends together and crimp to a wire. Make a hole from the bowl through to the electronics box and pass the wire though. The braid can then be attached to the bowl with hot glue.

Strip the ends of the wire and wrap around an M2.5 bolt, and twist the end to form a loop. Form a loop on the end of each resistor. Bolt one LED anode wire to the D0 connection on the Gemma, and the other anode wire to D2. The cathode wires should be bolted to the 0V connector. Bolt the sensor wire to the A0 connection.

BRAIN POWER

The Gemma M0 uses an ATSAMD21E18 processor, and this is powerful enough to run a subset of the Python language called CircuitPython. Combined with the provided libraries, this means we can build this kind of project with just a few lines of code.

```
import time
import board
import touchio
from digitalio import DigitalInOut, Direction
```

We start with instructions to use the libraries for time, board, and touchio. For the digitalio import instruction, we've just picked the items we need.

Above ◈
LEDs only work if the current flows from anode to cathode

Below ◰
Schematic

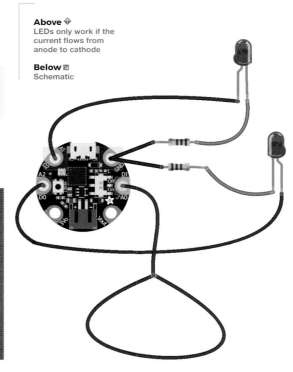

CAPACITIVE SENSING

The capacitive sensors on the Gemma work by charging and discharging the output pins and measuring the voltage. The pins have a small amount of capacitance and the microcontroller is sensitive enough to measure this. When a person is very near to the sensor, they affect the circuit by acting as a second capacitor. This means that when the microcontroller charges and discharges the pin, it will see a change in behaviour and know that it has been touched.

This can be a useful technique to simplify the code you write later.

```
touch_pad = board.A0
touch = touchio.TouchIn(touch_pad)
led1 = DigitalInOut(board.D0)
led1.direction = Direction.OUTPUT
led2 = DigitalInOut(board.D2)
led2.direction = Direction.OUTPUT
```

The pins need to be configured before we can use them. The A0 pin is configured for touch, D0 and D2 are configured as digital outputs. Note how `Direction` and `DigitalInOut` don't need a prefix as we imported those specifically above.

```
def detected():
        for i in range(3):
        led1.value = True
        led2.value = True
        time.sleep(0.5)
        led1.value = False
        led2.value = False
        time.sleep(0.2)
```

This code defines a function we can call when the skull is triggered. It flashes the LEDs three times, with a short delay between each flash.

```
detected()
while true:
```

```
    if touch.value:
        detected()
    time.sleep(0.05)
```

To check that the lights are working, the code calls the **detected** function when the Gemma first powers. The **while** statement creates a loop that will keep running until the power is turned off. It checks the touch sensor to see if it has been touched. If it has, then the code calls the function defined earlier to flash the lights.

Upload the code to the Gemma and remove the USB connector. Fit the batteries to the battery pack, and plug into the Gemma. Test everything is working, then fit the electronics into the back of the skull. Add some treasure to the bowl, e.g. sweets. Your monkey skull bowl is now ready for use. □

OTHER IDEAS

- Paint your skull in candy skull style for a 'Day of the Dead' look.

- Use PWM to make the eyes fade up and down, rather than flash.

- Use servos to make your skull move.

- Swap out the Gemma with a Circuit Playground Express board, and add some audio effects to your project.

QUICK TIP

On the Gemma M0, like many microcontroller boards, the pins are shared and can do different tasks.

Left ◈
Bolted connections using washers and 2.5 mm bolts

Sorting toys the robot way

Make picking through Lego a thing of the past

Above ⬆
The sensor fits into the design with a little hot glue, and the pins hold the brick in place whilst it's being analysed

Every home containing children (of whatever age) undoubtedly contains boxes upon boxes of unsorted construction toys – be it Lego, Hama Beads, Mega Bloks, Stickle Bricks, or any other assortment of multicoloured plastic. When eventually someone decides to sort 'that box' out, you might go for shape or size to categorise the thousands of tiny bits and pieces, but the more rainbow-obsessed among us might choose colour instead.

But sorting teeny plastic bits by hand is pretty laborious – it's 2019, we have robots! Surely they should be assigned this arduous task to save our fingers (and toes) from all those pointy bits of plastic?

This categorising contraption sees the bricks start in a queue at the top of the machine. One by one, they are inspected by a colour sensor, and a series of servo motors control a series of gates to determine the brick's path. Each one of these paths culminates in a pile of bricks of a certain colour. A Raspberry Pi is used to coordinate the sorting, running code written in Python as the brains behind the operation.

AN ALTERNATIVE APPROACH

Another option for colour sorting is to use a Raspberry Pi camera, along with the Python variant of computer vision library OpenCV. Sometimes known for being arduous to install, OpenCV can provide the means for very powerful image processing, including shape and colour recognition. You can find out more about OpenCV at **opencv.org**.

The build will vary a bit, depending on what exactly you want to sort – we went for the ubiquitous Lego 2×2 blocks as a starting point, but it should be easy to modify the CAD files for whatever you like.

SAVING YOUR FEET

The first step is to solder up all the circuit boards you need, and then connect them all up. Start with adding

Archie Roques

🐦 @archieroques

By day a humble school DT Technician, by night a hardware engineer, Norwich Hackspacer, and general projects man. Also blogs at **roques.xyz**.

Left ◈
This ingenious paper clip pin-release system was the idea of a fellow hackspacer. Collaboration for the win!

a 2 × 20-pin 0.1″ header to your Raspberry Pi if you're using a Pi Zero (we used a Pi 3B for ease, but this project will work with any version of the popular microcomputer). Then add a five-pin header to the BH1745 sensor, and solder up the first eight servo

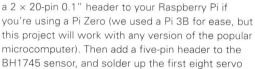

> Sorting teeny plastic bits by hand is pretty laborious – **it's 2019, we have robots!**

positions on the servo driver (numbered from zero to seven). It's also advisable to solder headers to either the left or right-hand row of pins on the servo driver. It's worth pointing out here that you could use any other servo-driving HAT, pHAT or add-on board, and any other RGB colour sensor – these are just the two we had to hand.

The next step is to wire everything up. Both the servo driver and the brightness sensor use I²C – a communication protocol for ICs – so they need connecting to the same two pins on the Raspberry Pi. However, I²C is a communications protocol that supports many devices on the same connections, so we just need to make sure everything can connect.

For this we used a Pimoroni Pico HAT Hacker, with some right-angled pins soldered underneath and some straight ones on top. It worked well for prototyping, but you could also solder wires directly between the circuit boards if you so choose.

You'll need to connect the following pins:
- 5 V on the Raspberry Pi to V+ on the servo driver (this powers the motors themselves)
- 3.3 V on the Raspberry Pi to VCC on the servo driver (this powers the driver chip)
- SCL on the Raspberry Pi to SCL on the servo driver and SCL on the BH1745 sensor
- SDA on the Raspberry Pi to SDA on the servo driver and SDA on the BH1745 sensor
- Ground on the Raspberry Pi to GND on the sensor and driver boards

Once you've done that, grab an SD card with a fresh copy of the Raspbian operating system on it and boot up the Raspberry Pi. You're best off using a beefy power supply with your Pi (we recommend the official one) as you'll be using it to power motors, and motors like lots of power.

When the Pi is booted up, connect it to the internet before starting to install the libraries required. The add-on boards have their own tutorials (the BH1745 here: **hsmag.cc/wwlhuY** and the servo driver here: **hsmag.cc/ullqcj**), which will →

YOU'LL NEED

◈ **Raspberry Pi** (any model)

◈ **Power supply**

◈ **Screen**

◈ **SD card**

◈ **Mouse and keyboard**

◈ **Servo driver** (or HAT)

◈ **BH1745 RGB colour sensor**

◈ **Micro servo motors**

◈ **600×400 mm sheet of 3 mm acrylic**

◈ **Jumper wires**

you've done all the installation and setup, reboot your Pi and then we'll be ready to rock.

Now it's time for some testing! To test if the colour sensor is running correctly, open up a new Python 3 window and add the following code:

```python
#import the libraries we'll use
import time
from bh1745 import BH1745

#set up the sensor
colourSensor = BH1745()
colourSensor.setup()
colourSensor.set_leds(1)

#print the colour every second
while True:
    print(bh1745.get_rgbc_raw())
    time.sleep(1)
```

SEEING IN COLOUR

When you run it, you should get readings from the sensor every second. If you hold up a red, green, or blue object to the sensor, you should notice the number in the first, second, or third column increase respectively. The last column is the overall brightness, and we won't be using that in our project so don't worry about it.

Next, it's time to test the servo driver. Plug in a motor to the leftmost socket on the board, with the data wire at the top (this is usually orange or white, but if you're unsure, your servo vendor should be able to let you know). Again, open up a fresh Python 3 window, and add the following code:

be useful if you're looking to take this project in a different direction. In essence, you need to run these commands:

```
sudo pip3 install bh1745
```

for the BH1745, and…

```
sudo pip3 install --upgrade setuptools
sudo pip3 install adafruit-blinka
sudo pip3 install adafruit-circuitpython-pca9685

sudo pip3 install adafruit-circuitpython-motor
```

for the servo motors.

You'll need to enable I²C support if you haven't already. To do this, run **sudo raspi-config**, go into interfacing options, and select 'Enable I2C'. After

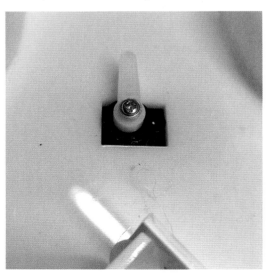

```
#import the libraries we need
from board import SCL, SDA
import busio, time
from adafruit_pca9685 import PCA9685
from adafruit_motor import servo
#initialise I2C communication
i2c = busio.I2C(SCL, SDA)

#set up the servo driver
pca = PCA9685(i2c)
pca.frequency = 50

#create a servo object
testServo = servo.Servo(pca.channels[0],min_
pulse=1000, max_pulse=2000)

#move the servo back and forward in a loop
while True:
    testServo.angle = 0
    time.sleep(1)
    testServo.angle = 180
    time.sleep(1)
```

This code should move the servo from its most clockwise position through to its most anti-clockwise position. It's helpful to add one of the little bits of plastic that come with the servos (known as horns) to tell where the motor is – these positions should be opposite each other; the servo should be travelling 180 degrees. If it's not, adjust the `min_pulse` and `max_pulse` values until it does – these can be any value from 600–2400.

Once these are all set up, then it's time to build the sorter.

We used cardboard for our prototype machine, and then laser-cut a super-accurate final model out of acrylic. You can use any 3 mm material of your choosing – or something thinner or thicker (if you're willing to resize the holes or get creative with the hot glue). You can download our CAD files (along with all the software) at **hsmag.cc/sgYfti**. You could also

GOING **FURTHER**

We built our sorter to sort only four colours of bricks – but (perhaps with the aid of a more accurate colour sensor, such as the AS7262) it'd be trivial to sort more colours. You could even have two stages of running – an initial sort to get the gist of the bricks' hue, followed by more detailed scrutinising. You could easily modify this build for marbles, Stickle Bricks, or Polo Fruits – the possibilities are limitless!

make this project with more traditional methods, if you'd rather.

LASERS! PEW! PEW!
The settings you'll need will depend on your laser cutter, and don't forget to make sure your ventilation is on and your material is laser-safe. The design will fit on a 500×700 mm sheet of acrylic, but you can dismantle it to fit your cutter.

 Waving them in front of a heat gun **for a few minutes will make the acrylic pliable**

Then, you'll need to fit your parts together. Depending on how thick your material actually is (which, as we found out, is quite often different from the stated width), you may need to use a bit of hot glue to aid your construction.

Some sections need to be bent for the corners. Waving them in front of a heat gun for a few minutes will make the acrylic pliable (and hot – make sure you wear heatproof gloves). Then you can fit it to the holes and hold it there until it cools. →

QUICK TIP
The servo control board has a space for a big capacitor to accommodate peaks of high current – we recommend it if you're using all but a few servos.

Below ◈
Our designs are for 3mm material. You will need to adjust them if using a different thickness

TUTORIAL ▬▬▬▬▬▬▬▬▬▬▬▬▬▬▬▬▬▬▬▬▬▬▬

QUICK TIP

All the code in this project is in Python 3 – if you use Python 2.7, some bits definitely won't work!

The sections are designed for micro servo motors to snap into the rectangular holes – though they need a bit of pressure and you might need to fettle the holes a little to get a good fit. There are two holes near the top of the chute, designed for a pin mechanism to release the Lego brick. The advantage of this is that it only releases one brick at once – as the front pin is

> **"** The sections are designed for servo motors to snap into the holes – **though they need a bit of pressure "**

down, the back one is up, keeping the next brick-in-line ready and waiting, and doesn't get deployed too early. We tried using various gates at first, but this often led to an avalanche of bricks at once. Bend some bits of paper clip into the holes, then glue the servo in place with some spare chunks of acrylic. Then you can run the trial code and trim the paper clip pieces to fit.

The last step is to attach the servo horns (the little arms that fit onto the rotating axis). A servo only moves 180°, so you'll need to modify the code above to set the servo to 90° and then fit the arm pointing to the middle of the chute so that it will be able to turn to either side to let the bricks past.

Then you'll need to run the code (available at **hsmag.cc/XuNDTV**) and test it out! One thing we found was that the sensor often didn't pick up the correct colour of bricks, because our white acrylic was too reflective. A coat of pound-shop black spray paint soon fixed that, and it worked fine after that. You can also change the criteria and variables in the code to reflect your lighting conditions etc.

And there you have it, a sorter for all your small coloured feet-stabbing building blocks! ☐

Below ◈
The pin mechanism for releasing one brick at a time

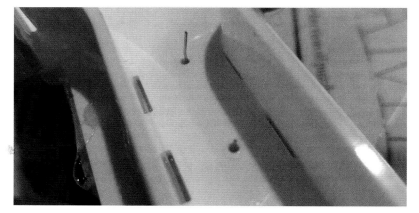

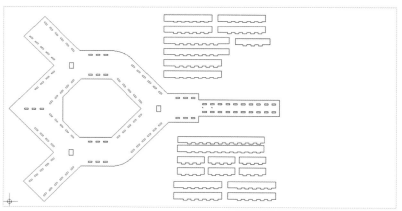

Above ◈
The contraption as a whole turned out to be quite large and took up the whole of our build table

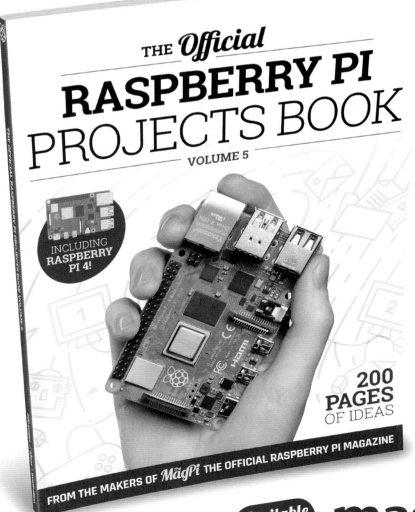

REGULAR

Black hole M87

3D-printed representation of the first black hole photograph

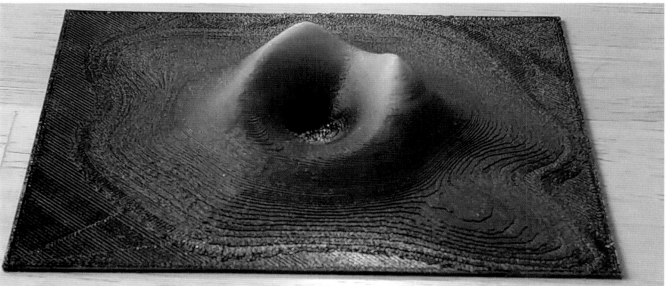

T he science world got a treat in April 2019 with the publication of the first ever photograph of a black hole. Well, kind of – black holes suck in light, so you can't actually take a picture of one directly. The image was instead generated by tons of data gathered by the Event Horizon Telescope array, made up of individual radio telescopes around the world. This brilliant original piece of work by Dean Segovis takes that photo and takes it to (as the Prodigy would say) another dimension.

There's a feature in Cura that enables you to extrude a colour upwards out of a 2D image, as long as it's high-contrast, making it 3D. You can change the dimensions, and the smoothing of the image – based on the video that Dean has put up on his YouTube channel, the defaults look pretty good, with a touch extra smoothing.

From Cura, you can export to G-code or an SDL file for printing. The model was then airbrushed to match the colours of the photo. Bingo – a black hole for display on the wall. It's slightly smaller than the real thing at 120 mm square (black hole M87 is 6.5 billion times the mass of the sun), but it is just as good a talking point. ◻

↗ **hsmag.cc/ayCuVt**

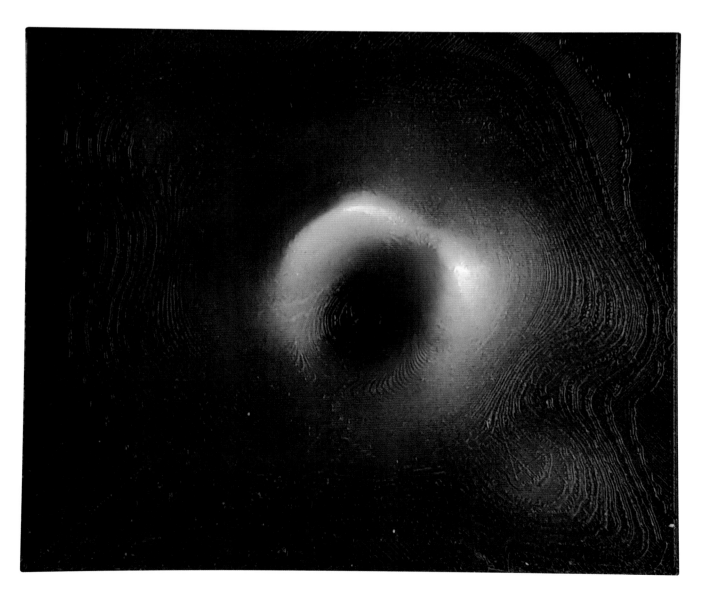

WEEKEND PROJECTS

HACK | MAKE | BUILD | CREATE

Got a whole weekend spare to do some making? Get stuck into these more advanced projects

PG
76

BUILD YOUR FIRST ROCKET

Make a model rocket and launch it into the sky

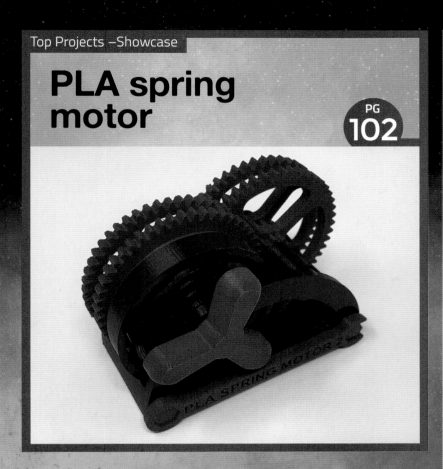

Build Your
FIRST
ROCKET

Reach for the stars

HackSpace
TECHNOLOGY IN YOUR HANDS

Ben Everard

🐦 @ben_everard

Ben loves cutting stuff, any stuff. There's no longer a shelf to store these tools on (it's now two shelves), and the door's in danger.

I'm heading to Midland Sky for my first taste of model rocketry. Just off the M42 in the West Midlands, we follow a farm track to a strip of grass alongside the field that serves as car park, camp-ground, and launch site. Occasional clouds push across the post-harvest fields and a group of rocketeers gather around, readying their crafts.

There are two groups of launch pads – one for High Power Rocketry (HPR) and a smaller one for smaller rockets, closer to the group of spectators. In both cases, the launch pads look like poles, supported at the bottom and poking at the sky – although later, when I see them up close, I find that the model pad is thin metal poles, while the larger ones are slotted shafts (the rockets have pins that slot into these).

A group of people approach and fill the pads with rockets, then retreat to a safe distance. Paul Carter, the Range Safety Officer (RSO), summons each rocketeer in turn to launch. Checking that there's nothing overhead (we're between two airports, and adjacent to a microlight airstrip), he issues a countdown.

Five
Four
Three
Two
One

VVVVWOOOOSH!

The engine lights and, leaving a trail of smoke, the craft flies skywards. Even though I'm not involved in the rocket, the roar, the smoke, and the flash of flames induce a tingle of exhilaration in my veins. The closest thing that I can describe it to is a penalty shoot-out. Although it's non-competitive, there's the same build-up, the same explosive release, and the same sudden success-or-failure.

Most engines burn for only a few seconds, but that's enough to send them hundreds, or thousands, of feet into the air (the club uses feet rather than metres).

The crowd of spectators (mostly, but not entirely, fellow rocketeers) crane their necks upwards. Even the larger rockets (over two metres in length), can be hard to spot at high altitude,

and having more eyes means it's more likely to be tracked.

Once it's reached its high point (the 'apogee' in rocketing terminology), it starts to fall. The simple, smaller rockets have a small charge that detonates a few seconds after the motor; this pops out a parachute and it glides down safely. However, the larger and higher-flying rockets don't use this automatic method, as it would deploy the parachute high up – this means that there'd be a long drift as the wind carried it across the field, and possibly into a neighbouring farmhouse. These use an altimeter to automatically release the parachute at a preset height.

As the rockets glide safely to earth, the crowd applauds – both in congratulations of a successful launch, and in appreciation for the show that's been put on for them.

Once all the rockets have launched, the rocketeers set off across the field to retrieve their craft, and a new group of people begin setting up their rockets.

At least, this is how the rockets should launch. A few go wrong. Mostly, they fail to ignite the motor – usually due to a duff igniter. A few spin spectacularly, or fail to deploy their parachutes, and end up coming back in more pieces than intended. →

Above ◈
Once you've learned the basics, you can improvise with different materials

Below ◈
A line-up of model rockets ready for launch

The noise, smoke, the rockets' raw power, coupled with the nerves of it not firing, combine to make it an exhilarating experience. I can't claim to be a rocketeer, but I have built and launched a couple of rockets! →

Build your first rocket
—
FEATURE

Design YOUR OWN

Personalise your rocket using specialist software

Jo Hinchliffe

🐦 @concreted0g

Jo Hinchliffe is a contributor to the Libre Space Foundation and sits on the UK Rocketry Association council. He loves designing and scratch-building both model and high-power rockets, and releases the designs and components as open-source.

O nce you've tried a rocket kit, you might want to move up to something a little more personalised. Designing your own rocket can be as simple or complex as you like. There are a few bits you need to take into consideration, such as the estimated height and the stability. Beyond this, you can complicate things with multiple motors, or by making scale models of other craft, or you can keep things simple. You can go for ultimate power, or you can keep things at a smaller scale. However you do it though, it's easier with some design software to help you. Let's take a look at how to design a rocket using OpenRocket, one of the most popular bits of design software for hobbyists.

OpenRocket helps you understand how different parts will fit together and how these will affect the flight characteristics. You can even put the rocket through a simulation that estimates how high the rocket will go. Let's get started on our rocket.

First, we'll need to install the relevant software. OpenRocket runs on Java, so you should be able to run it on Windows, Mac, or Linux, provided you've got the runtime environment installed (you can get this from **java.com/en/download** if you don't have it already). With this, you can download OpenRocket from **openrocket.info**. It's a JAR file, so there's no need to install it. Just click on the downloaded file to start.

When you first start OpenRocket, it'll start a new rocket project for you and you'll be greeted with a screen similar to **Figure 1**.

Traditionally, rockets have a pointy bit at the front, a long straight bit in the middle, and some fins at the

Right ◈
The Open Development rocket on the launch pad at Midland Sky

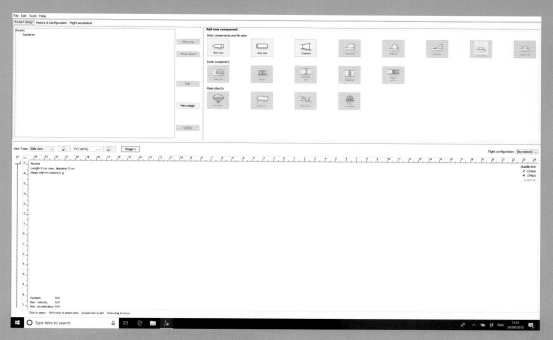

Figure 1 ◈
The layout of OpenRocket has all the key information split up into three panes

back, and ours will be no different. Let's start with the pointy bit at the front (technically, this is known as a nose cone).

Click on Nose Cone in Add New Components, and you'll see a roughly triangular shape appear in the side-view panel. You'll also get a dialog box that lets you customise this. Enter the following options:

- **Nose cone shape:** Ellipsoid
- **Nose cone length:** 8 cm
- **Base diameter:** 3 cm
- **Wall thickness:** 0.2 cm
- **Material:** PVC

We'll also need to add a shoulder that lets it fit into the main body tube. Click the Shoulder tab and enter the following:

- **Diameter:** 2.6 cm
- **Length:** 2 cm

Click Close to get back to our design.

We'll now create the main part of our rocket. Click on Body Tube, and enter the following in the dialog:

- **Length:** 45 cm
- **Outer diameter:** 3 cm
- **Set material:** Cardboard

The inner diameter should automatically set to 2.6 cm, giving a wall thickness of 0.2 cm. Double-check that this is correct, then click Close to return to our rocket.

The final part of our rocket outline is the fins at the back. To add this, click the Trapezoidal icon. You will need to have the body tube highlighted in the plan view for the fin set icons to be available to select. Either click on the body tube itself in the plan view area, or click its name in the component tree area.

In the dialog box, set the following:

- **Number of fins:** 3
- **Root chord:** 4 cm
- **Tip chord:** 4 cm
- **Height:** 2 cm
- **Sweep:** 1.5 cm
- **Sweep angle:** 36.9
- **Fin cross section:** Rounded
- **Thickness:** 0.3 cm
- **Material:** Birch

Then click Close.

You should see now that the centre of pressure (CP) symbol (red dot inside red circle) is behind (toward the fin end) the centre of gravity (CG) symbol (blue quartered circle). The top right-hand corner of the plan box contains some text about their positions. The 'Cal' number is to do with stability. It is the number of calibres (diameter of the body tube) behind where the CG the CP is. We ideally aim for the CP to be between 1 and 2 calibres behind the CG. Over 2 calibres is OK, but is classed as 'overstable' and may turn more into any prevailing wind, resulting in a ballistic flight. At the moment, the stability is 2.26 Cal, but this will change as we continue to alter the design. →

EXAMPLE ROCKETS

We've built a simple rocket, but you can browse the other example rockets for some other designs. Click on File, then Examples, to see the list of available designs. There are some showing more advanced capabilities such as clustered motors and multiple stages. You can tweak these, or build new rockets based on them, but make sure that you match your designs with your skill as a rocket builder – don't jump straight to advanced builds without gaining the necessary skills first. Rocketry clubs are a great place to get guidance on how to proceed.

We've got the outside of the rocket designed. It's time to move inside and look at how we'll secure the motor. With the Body Tube highlighted, click the Inner Tube icon and, in the dialog box, set the following parameters:

- **Outer diameter:** 1.9 cm
- **Inner diameter:** 1.8 cm
- **Wall thickness:** 0.05 cm
- **Length:** 7 cm
- **Material:** Cardboard

In order to make sure that OpenRocket knows where to put our motors, we need to let it know that this is where we want to put one. In the Motor tab, check 'this is a motor mount', then click Close.

The inner tube is just floating around inside the rocket at the moment; we need to add some bits to secure it to the main tube. With the Inner Tube (motor mount tube) highlighted (you'll need to select it in the component list), click the Centring ring icon, and enter the following in the dialog:

OPEN-SOURCE ROCKETRY

If you're looking to get started at designing your own rocket, an open design is a great place to begin.

The Open Boat Tail rocket takes low-power 18 mm engines, so is great for people getting started in rocketry. You can find the designs at **hsmag.cc/AOudlm**.

If you're looking for something with a little more 'oomph' (mid- to high-powered rocketry), then the Open Development Rocket is a great place to start. It's a rocket designed to be made using a 3D printer and laser cutter (as well as a few commonly available parts, such as cardboard tube). It takes a 38 mm motor in the mid-to-high power range (G to I).

On its maiden flight, the ODR reached 3255 feet, but deployed the parachute a little bit too early (possibly due to a loose nose cone shaking as it fell, rather than waiting until it had reached the preprogrammed deployment altitude), so if you base your design on this rocket, you may need to make adjustments to take this into account. You can find the designs at **hsmag.cc/TFEFJa**.

In both cases, you'll find everything you need to 3D-print and laser-cut your rocket. Couple this with some easy-to-find hardware (such as some cardboard tube and an altimeter), and you'll be ready to fly.

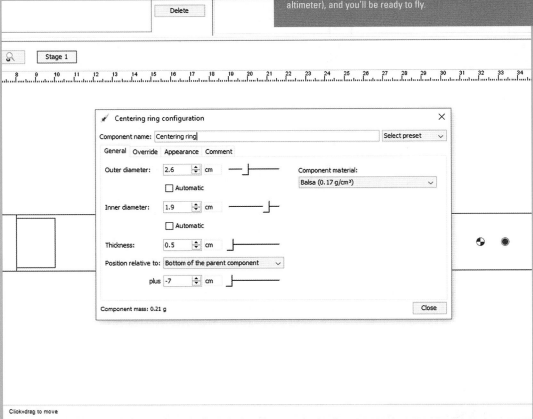

Delete

Stage 1

Centering ring configuration ✕

Component name: Centering ring   Select preset ⌄

General | Override | Appearance | Comment

Outer diameter: 2.6 ⌄ cm ———   Component material:
☐ Automatic   Balsa (0.17 g/cm³) ⌄
Inner diameter: 1.9 ⌄ cm ———
☐ Automatic
Thickness: 0.5 ⌄ cm ———
Position relative to: Bottom of the parent component ⌄
plus -7 ⌄ cm ———

Component mass: 0.21 g   Close

Right ⬀
Selecting the correct material helps OpenRocket understand how the rocket will behave in flight

Click+drag to move

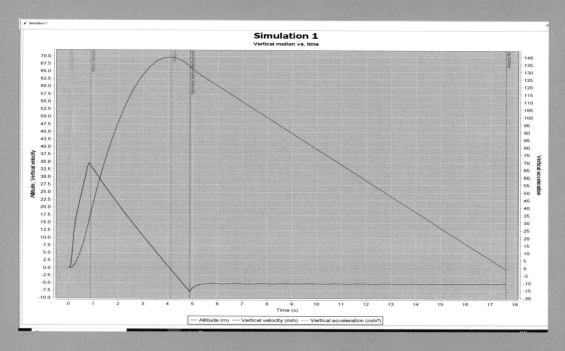

Simulation 1
Vertical motion vs. time

- **Outer diameter:** 2.6 cm
- **Inner diameter:** 1.9 cm
- **Thickness:** 0.5 mm
- Set 'position relative to Bottom of the parent component' to -7 cm
- **Material:** Balsa

Repeat this to add a second ring with the same values, except the 'position relative to Bottom of the parent component' as 0. This will put a mount at either end of the motor tube to fully secure it to the main body.

We're almost there now. The only thing left is to ensure that our rocket can come safely down to earth. Highlight the Body Tube and click the Parachute icon. Then enter the following in the dialog:

- **Diameter:** 30 cm
- **Position 'top of the parent component':** 3 cm

Now is the time to save your work. Click File, then Save As, and give your work a unique file name if you haven't done it already.

READY FOR LAUNCH
We've got our rocket designed, so now it's time to see what it'll do in flight. Click to highlight Inner tube/motor tube, and then click the 'Motors and configuration' tab in the tabs area. Click New

Configuration, then double-click on None in the Inner tube column.

There are all kinds of ways you can constrain the lists of commercially available motors. For example, you could set the software to search for an individual motor manufacturer or for the motor diameter. For our run-through we need to find an 'Estes B6-4' motor and select it, making sure to add a four-second delay in the motor selection dialog box, then click Close.

Now everything's set up, let's see how this flies! Click the Flight simulations tab from the main three tabs.

You should see that there is a single simulation listed that hasn't been run with your motor details in it. Highlight it, and then click the 'run simulation' button. The slots will fill with data. Then, click the Plot button and OK the next dialog, and you should get a flight simulation graph.

Congratulations! If you made it this far, you have designed and simulated a small, and not terribly efficient, rocket. Time for you to experiment and play and design something amazing. →

JOE BARNARD

Meet the maker attempting to land model rockets

Below ⌐
The electron rocket that Joe's been attempting to land vertically. Note that the fins are on the top to help stability when coming down

Most model rockets follow the basic flight process that's been unchanged since the Space Race. A motor accelerates the rocket upwards. At some point, the engine stops burning – the rocket then slows, and falls to earth. A parachute is deployed and glides the craft down as safely as possible. However, since SpaceX pioneered vertical landing (where a second burn of the engines slows the rocket down, allowing the rocket to land vertically on the pad), hobbyist rocketeers have been keen to emulate this feat.

Joe Barnard is attempting to do this using solid-fuel rockets, and regularly uploading videos on his progress to YouTube at: **hsmag.cc/pZALEA**. These engines are cheap and simple to use, but don't have any control over the thrust. Once they're lit, they

keep burning until they're spent, unlike hybrid or liquid engines that can vary their thrust. This means that it has to be lit at exactly the right moment (or at least within about 20 ms of the right moment) in order to land smoothly.

PRACTICE MAKES PERFECT
"It's not perfect, but as soon as you start adding actual throttle control, everything gets about 1000 times harder because then you need plumbing, you need much stricter safety procedures. If you're willing to accept that – just by luck – some of your flights aren't going to work, then you can get really close with solids."

While the amount of thrust can't change, the direction can. Joe developed a thrust vector control system that can angle the motor in order to keep the rocket upright as it takes off and returns to earth.

In order to simplify the problem, Joe's currently working on just the landing, by dropping rockets from drones.

"I think I'm getting pretty close to getting it. I don't know if it's going to be a couple more tests, or if it's going to be another three years. We'll see."

Joe's work is unashamedly inspired by SpaceX. In fact, Joe first started working on thrust vector in order to hopefully impress SpaceX enough to get a job there. He's no longer working towards this aim, but you can still see this inspiration in some of his rockets, including the 1/48th scale model of a Falcon Heavy that he's working on, complete with thrust vectoring in the boosters and second stage.

Looking to recreate Joe's style of rocket? He sells the Signal thrust vectoring system, which comprises a flight computer and 3D-printed mount that uses two 9 g servos to control the motor. He's also publishing a series of videos, entitled 'Landing Model Rockets', that go through the technicalities of doing exactly this. →

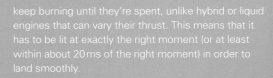

Left ⬦
An evening launch
of Joe's scale model
Falcon Heavy

Below ⬦
Using thrust
vectoring allows
slower launches,
which are more
realistic compared
to full-sized rockets

Credits
Joe Barnard,
BPS.Space

COPENHAGEN SUBORBITALS

Inside the amateur space programme run by hobbyists

Right ⬦
The Heat 1X leaving the launch pad. At the top you can see the capsule for an astronaut

Credits
Copenhagen Suborbitals – Thomas Pedersen

Left ◈
**The Nexø II
rocket being
safely recovered
from the Baltic Sea**

Credits
Copenhagen
Suborbitals –
Mads Stenfatt

Copenhagen Suborbitals (CS) is
an amateur space programme
aiming to put a person in space.
That's an ambitious target – you
need a powerful rocket, a reliable
launching system, and a way of
getting an astronaut safely back to Earth. This has
obviously all been done before, but only by huge
space programmes run by the world's largest nations.
Replicating this achievement with a completely
volunteer staff of hobbyists is a massive challenge,
but it's one that CS seems determined to meet. The
CS team have already been working on this for seven
years, and have launched seven rockets, each one
getting them a little closer to their goal.

CS's first rocket, HEAT 1X stood 9.38 metres
tall and weighed 1680 kg on the launch pad – an
enormous rocket by amateur standards. As a
testament to the team's ambition to launch a person
into space, the HEAT 1X included a capsule with
a life-size crash test dummy. It was powered by a
hybrid engine which used a liquid oxidiser (nitrous
oxide) and a solid fuel (polyurethane).

Unfortunately, the rocket started to pitch to one side
immediately after launch, and ended up at an angle
30 degrees off horizontal, and had to be manually
shut down (via the radio link). A post-mission analysis
concluded that a misaligned engine nozzle caused the
problem. As HEAT 1X was passively stabilised, there
was no process to correct this problem in flight.

Since then, CS has been launching smaller rockets
and building up the technology bit by bit. The most
recent rocket, Nexø II, took off from CS's floating
launch platform in the Baltic Sea on 4 August 2018.

Powered by liquid ethanol and oxygen, with thrust
vectoring to able it to adjust its flight, this is a far more
advanced rocket than HEAT 1X. It reached an altitude
of 6500 metres, before returning to Earth successfully
via a parachute and a splashdown. Although this was
far too small to carry a person (6.7 metres long, 30 cm
wide, and weighing 292 kg at take-off), this is the first
CS rocket to achieve a landing soft enough that a
human could have survived. This launch wasn't without
problems, though, failing to reach its intended altitude
of 12,000 metres. The team are still investigating why
this happened.

BIGGER AND BETTER

The final goal of CS lies in the Spica rocket, which
will stand 13 metres tall, 95.5 cm wide, and weigh
4000 kg. This will be powered by a liquid-fuel engine
similar to those in the Nexø II, but scaled up to provide
100 kN of thrust (compared to 5 kN on the most recent
launch). Hopefully, the Spica rocket will be able to
reach an altitude of over 100 km above Earth (known
as the Kármán line, and generally accepted as the limit
of space) and deploy a capsule which will contain an
astronaut who will then come safely down to sea.

If you're interested in what it takes to build and
launch rockets of this scale, CS has an informative set
of videos that take you through the things the team are
building and the problems they encounter. You can take
a look at **hsmag.cc/DJOviq**.

CS is financed by supporters who either sign up
to contribute monthly (in a Patreon-like fashion), or
buy merchandise. If you're keen to see them reach
their goal, you can help sponsor the mission on their
website: **hsmag.cc/XSGpAt**. ▢

Polyphonic digital
synthesizer Part one

Build a full-featured polyphonic digital synthesizer in our two-part guide

Matt Bradshaw

mattbradshawdesign.com

Matt Bradshaw is a
programmer, maker, and
musician from Oxford. He
likes to build instruments
to play with his band,
Robot Swans. You can
find more of his projects at
mattbradshawdesign.com

Analogue synthesizers have made
a big comeback in the last few
years, but building a synth that
can play multiple notes (i.e.
chords) using only analogue
circuitry is a big challenge. In
this tutorial, you will see how to build a versatile
synthesizer with a 'patchable' signal chain, but where
the sound is generated digitally by code that you can
write yourself. This is a two-part tutorial, but even
by the end of part one you'll already be able to make
some great sounds.

GOING DIGITAL

Modular synthesizers are awesome. They let you
create your own signal chain by plugging cables into
different points in the circuit, giving you the freedom
to create any sound you can imagine. A true modular
synth is basically a box which you populate with

Above ◈
This design combines aspects of digital and analogue synthesizers,
to give you a versatile, but cheap-to-make, instrument

individual modules that you can either buy or build
(see the excellent tutorial from HackSpace magazine
issue 14 for an example). Some modules generate
signals, while other modules take a signal and change
it in some way.

This tutorial will show you how to create a
miniature, digital version of an analogue modular
synthesizer. The process of 'patching' different
signals into each other will be done on the breadboard
with jumper wires, and that information will be
processed by the Teensy microcontroller.

Firstly, we need to set up our Teensy 3.2, which is
a bit like an Arduino but powerful enough to process
audio. When you buy a Teensy, it usually comes

Above ◈
The extra-long headers are soldered underneath the Teensy and the audio board

Above ◈
Here's how the Teensy and audio board should look on the breadboard – make sure that the pin numbers line up

YOU'LL NEED

◈ **Teensy 3**

◈ **Teensy audio adapter board**

◈ **4 × 14-pin stackable male/ female headers** (2 kits)

◈ **2 × breadboards**

◈ **Jumper wires**

◈ **2 × rotary potentiometers** (10 kΩ, linear)

◈ **5 × 4051 multiplexer chips**

◈ **8 × tactile buttons**

◈ **LED**

◈ **6N139 optocoupler chip**

◈ **MIDI socket**

◈ **Capacitor** (0.1 µF)

◈ **Resistors** (various)

◈ **USB micro cable**

◈ **Soldering equipment**

◈ **Headphones**

◈ **Computer**

without any headers, so you'll have to do a bit of soldering. The audio board, which sits either directly above or below the Teensy, also requires soldering. Using stackable headers is a good idea (female headers with long male pins on the other side), as these will make both the Teensy and the audio board compatible with a breadboard.

Once your Teensy is ready, download the Teensyduino software from **hsmag.cc/aRWmgD**,

> **The Teensy has its own library of code** for adding audio to projects

and try running an example audio sketch, such as File > Examples > Audio > Synthesis > PlaySynthMusic. You should then be able to hear music through the headphones jack.

START SMALL
Before we can connect lots of modules together, we should try a simple sketch to get the hang of writing code to produce audio. The Teensy has its own library of code for adding audio to projects, and it works a lot like a modular synthesizer.

For instance, in the **sine_wave** example sketch, an oscillator is connected to an output via two `AudioConnection` instances (one for each stereo channel), meaning that a sine wave is heard through the headphones. Download this sketch from **hsmag.cc/issue16**, and try it for yourself.

Our synth will consist of eight sockets, and we will need to know which sockets are connected to each other. For instance, if the oscillator socket is connected to the main output socket, the Teensy needs to be able to read this and then recreate the connection digitally, producing audio. On a 'real' synth, these sockets would be sturdy 3.5 mm connectors (basically headphone jacks), but for this synth we are simply going to use a row of breadboard sockets. For now, it doesn't really matter which socket corresponds to which input or output – we just want to know whether socket A is connected to socket B, and so on.

ONE THING AT A TIME
In order to test the connections between the sockets, we will use an integrated circuit called the 4051. This is an eight-channel multiplexer or demultiplexer; in layman's terms, eight 'things' are connected to the chip, and you can talk to them →

AVERTING THE **SPAGHETTI**

This synth, particularly once you complete part two, will involve a lot of wires in a relatively small space. If you use lots of standard-length jumper wires, you will quickly end up with an unmaintainable rat's nest (albeit a very pretty one). To alleviate this problem, it's worth making a batch of your own tiny jumper wires from single core wire, maybe 4 cm long each, with about 5 mm of insulation stripped away at each end.

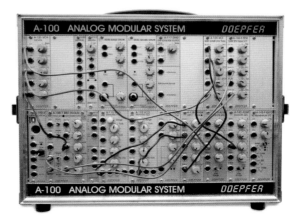

Left ◈
This is what a full modular synth looks like, with removable modules and patch cables

Polyphonic digital synthesizer (Part one)

Figure 1 ◈
The full breadboard
layout, with the audio
board omitted for
clarity. Make sure to
connect the channels
of the two 4051 chips
(see orange wires)

Figure 2 ◙
In the 'connection_
test' sketch, you
can check that
your circuit is
working correctly

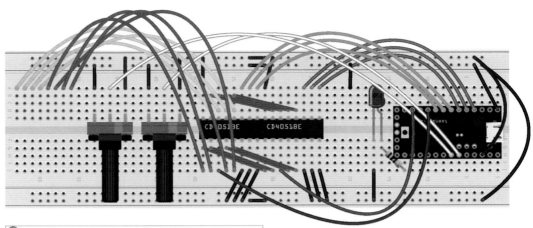

```
for(int b=0;b<8;b++) {
  setReadChannel(b);
  delayMicroseconds(10);
  if(a < b) {
    boolean connectionReading =
!digitalRead(CONNECTION_READ_PIN);
    if(connectionReading) {
      Serial.print(a);
      Serial.print(" is connected to ");
      Serial.print(b);
      Serial.print("\n"); }}}}
```

Upload the whole sketch to the Teensy and open
the serial monitor. Now try connecting two of the
sockets on the left end of the breadboard with a
jumper wire. If everything is working, the serial
monitor should report that a connection has been
detected (see **Figure 2**), and we're ready to move
onto the actual synth code.

THE INS AND OUTS

The 4051 gives us a maximum of eight sockets to
use, which are allocated as follows:

- Oscillator output #1 (square wave)
- Oscillator output #2 (sawtooth wave)
- Oscillator frequency modulation input
- Low-frequency oscillator output
- Filter input
- Filter modulation input
- Filter output
- Main output stage

These sockets are worth explaining in a bit more
detail, especially if you're not that familiar with
synthesizers. The two oscillator outputs are simply
tones with slightly different sounds (the sawtooth
is a bit more 'buzzy'). The oscillator modulation
input changes the pitch of the oscillator, meaning
that when you connect the low-frequency oscillator

one-at-a-time. Three pins are used to select which
'thing' you want to talk to (these are connected to the
Teensy), and eight pins are connected to the 'things'
(in our case, the sockets).

Start by building the breadboard circuit, as shown
in **Figure 1**. Notice that there are two 4051 chips,
both addressed separately, but with their channels
connected to common sockets. By using two 4051
chips in this way, you can send a test signal to each
channel in turn on the first chip, then listen for that
signal on each channel in turn on the second chip.
If a signal is sent to channel A on the first chip, for
example, and can be read on channel B of the second
chip, socket A must be connected to socket B.

To try this out, download the **connection_test**
sketch from **hsmag.cc/issue16**, and open it in the
Arduino IDE. You will see a nested **for** loop, with the
outer loop addressing the 'send' chip and the inner
loop addressing the 'read' chip.

```
for(int a=0;a<8;a++) {
  setSendChannel(a);
```

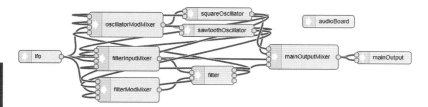

CHEAPSKATE **VERSION**

This project is a pretty cheap way into building your own synthesizer, but if you're willing to put in a bit more effort you could make it even cheaper. The audio board used in this tutorial is great, but there are less expensive alternatives. The PT8211 audio chip will give you 16-bit audio output for very little money, if you don't mind some very delicate soldering. Alternatively, you can get a lower-quality audio output direct from the Teensy via its DAC pin. Note that both of these options will require minor changes to the code.

(LFO) into it, you'll hear a tone that rises and falls like an ambulance siren. The filter is an effect which restricts certain frequencies while boosting others, and can also be modulated by the LFO. Finally, the main output stage represents the final part of the signal chain – you won't hear anything until you plug something into it. Don't worry if you don't understand all the ins and outs – once you start playing around with the synth, it should all start to make sense.

MAKE SOME NOISE

The easiest way to start writing audio code for a Teensy is to use the online 'Audio System Design Tool' at: **hsmag.cc/OiKbYH**. It's a simple drag-and-drop interface for connecting audio modules together, and it's definitely worth getting familiar with. For this synth, however, you can just copy and paste the code directly to make things a bit easier. Download the main sketch code from: **hsmag.cc/issue16**.

It's a good idea to look through the code to understand what's going on. The sketch basically combines the two simpler sketches from earlier, and adds a few extra features. It begins by declaring the various audio objects and how they are connected – this code was generated in the online design tool. Next, we declare an array of references to the four input mixer objects, so that we can easily reference them by number later on.

In the `setup()` function, we initialise the various input and output pins, and set some initial parameters for our audio objects – feel free to tweak these numbers to produce different sounds. The `loop` function works much the same way as in the earlier example sketch, but instead of sending a serial message when a connection is made (or broken), the volume of a relevant mixer channel is set to either one (for a connection) or zero (for no connection).

At the end of the loop, the LED is lit if a bad connection (input-to-input or output-to-output) is

detected. Unlike on an analogue synth, making bad connections won't cause any harm in this design, but it's useful to know. Finally, the two potentiometers' values are read and used to control the LFO frequency and main oscillator frequency. Feel free to change this section of the code to customise your synth, by making the knobs control other parameters.

The last job is to label the patching area on the left of the breadboard. Either use a fine pen, or print a label from your computer in a small font, and use Blu Tack or tape to affix the label to the breadboard. Now you can start playing!

Try connecting different outputs to different inputs and see what happens. Turn the knobs up and down to control the sound. The synth is capable of dirty bass drones and *Doctor Who*-esque effects, but if you want to get really musical, you'll have to wait for part two! □

QUICK TIP

If this synth has piqued your interest, try the free, open-source software 'VCV Rack', which is a virtual modular synth.

Below ◈
A sneak peek of what the synth will look like after part two, including a mini keyboard and MIDI input

NEXT **TIME**

In the second and final part of this tutorial (page 92), we'll be adding some features to really turn this project into a usable synthesizer. We'll be adding a second breadboard with a simple keyboard (allowing you to play melodies) and a MIDI input (allowing you to control the synth from another keyboard or a computer). We'll also double the number of connections you can make, and use a cunning trick to add polyphony to the synth, meaning you can play more complex music.

Polyphonic digital synthesizer **Part two**

The conclusion of our two-part guide to making a polyphonic digital synthesizer

Above ↖
The finished synth,
featuring 16 patch points,
two analogue controls, a
miniature keyboard, and
a MIDI input

L ast time, we built a digital synthesizer on a breadboard. It could make some fun noises, but it wasn't very useful for playing music. This time, we're going to rectify that by adding a simple keyboard, as well as a 'MIDI input' port so that you can control the synth with an external keyboard. We're also going to double the number of patch points so you can create more complex sounds. Finally, we're going to edit the code to allow the synth to play multiple notes simultaneously.

If you haven't already read part one, go back and start there – otherwise, let's get stuck in. We've already filled our first breadboard, so we need to add another one. We can leave a lot of our first synth in place: the Teensy, audio board, LED, resistor, and the two 4051 chips can remain untouched on the first breadboard. However, in order to make space for our awesome new features, you should remove the two potentiometers (and their wires), the row of eight

wires that connect the 'patch points' to the 4051 chips, and the label that showed what each patch point did.

The full new layout can be seen in **Figure 1** (overleaf). There is a 6N139 optocoupler (explained later) and three extra 4051 chips. Two of the 4051s perform the same function as in part one, detecting which patch points are connected to each other, but by adding another two chips, we are able to double the number of patch points.

The final (leftmost) 4051 chip acts as a multiplexer for the eight buttons of our mini-keyboard on the front breadboard. These eight buttons will play a simple major scale, although you can change this in the code if you'd prefer a more interesting set of notes.

The front breadboard also now contains the MIDI input, the 16 patch points, and the two potentiometers, so all of the 'hands-on' components (things you might want to access during a performance) are easily accessible.

WHAT'S NEW?

Before we assemble everything, let's look at what new 'modules' we're adding. The synth already has two oscillators, a low-frequency oscillator (LFO), and →

Matt Bradshaw

mattbradshawdesign.com

Matt Bradshaw is a programmer, maker, and musician from Oxford. He likes to build instruments to play with his band, Robot Swans. You can find more of his projects at **mattbradshawdesign.com**

Below ◈
Taking the synth for a test drive – playing with a proper keyboard via MIDI opens up a wider range of notes than the breadboard buttons

YOU'LL NEED

◈ **Teensy 3**

◈ **Teensy audio adapter board**

◈ **4 × 14-pin stackable male/ female headers** (2 kits)

◈ **2 × Breadboards**

◈ **Jumper wires**

◈ **2 × Rotary potentiometers** (10 kΩ, linear)

◈ **5 × 4051 multiplexer chips**

◈ **8 × Tactile buttons**

◈ **LED**

◈ **6N139 optocoupler chip**

◈ **MIDI socket**

◈ **Capacitor** (0.1 μF)

◈ **Resistors** (various)

◈ **Micro USB cable**

◈ **Soldering equipment**

◈ **Headphones**

◈ **Computer**

◈ **1N4148 diode**

Polynomic digital synthesizer (Part one)

Above ◈
Many MIDI devices
have three ports:
'in', 'out', and 'thru' –
make sure to connect
the MIDI out port of
your external device
to the MIDI in port on
the breadboard

WHAT CONNECTIONS ARE ALLOWED?

In part one, we briefly discussed the 'bad connection' LED, which lights up if you make a connection other than input-to-output. This is a useful feature for diagnosing why your patch might not be working (perhaps you accidentally connected an oscillator to the filter output instead of the input). However, there are some valid patches which will also trigger the LED. If, for instance, you connect both the square and sawtooth oscillators to the main output, the synth will happily mix the two signals, but the LED will illuminate. This is because, electrically, the two oscillator outputs are now connected to each other in a circuit. If you would like an interesting little programming challenge, you could extend the LED code to detect valid connections such as this and disregard them.

a filter. This time we will add an amplifier, an envelope generator, and a MIDI-to-CV converter.

Briefly, an amplifier takes an audio signal and changes its volume. If you feed the module a high control signal, the audio will be loud, while a low control signal will quieten the audio. You can therefore use this module to make an oscillator 'turn on' when you hit a key and 'turn off' when you release it. However, notes that just turn on and off suddenly are not very interesting, which is where the envelope generator comes in.

An envelope generator (EG) mimics the sound of an acoustic instrument. When triggered by an input control signal, often from a keyboard key being pressed, the EG outputs a control signal which, when connected to an amplifier or filter, can evoke the sound of a guitar, a violin, or a piano (depending on the settings).

Finally, a MIDI-to-CV converter takes a MIDI signal from an external keyboard and converts it to a CV (control voltage) signal. This module outputs a 'note' signal (which communicates the last note to have been pressed), and a 'gate' signal (which is simply high or low depending on whether a key is currently being pressed).

Don't worry if these descriptions are new to you – YouTube has plenty of videos detailing how different synth modules work if you'd like to learn more, and we've provided some patching examples to get you started.

WE WILL REBUILD
Now we've got an idea of the new modules, let's add some components. It makes sense to build the circuit step by step, so we can check for errors at each stage. Firstly, using **Figure 1** for reference, add the two potentiometers, as well as the two 4051 chips directly to the left of the existing ones, and wire them up as shown.

Remember that the original synth required two of these eight-channel chips to provide eight patch points: one chip sends a test signal while the other reads it. By adding another two chips, we can have 16 patch points.

You could patch directly between the chips but, like last time, it's a lot easier if we run a jumper wire from each patch point to a separate, labelled patching area. These wires are omitted on the breadboard diagram for clarity (there are already too many wires on there!), but there is a separate zoomed-in diagram (**Figure 2**) with the patch points labelled as follows:

A) LFO (out)
B) Sawtooth oscillator (out)
C) Square oscillator (out)
D) Filter (out)
E) Envelope generator (out)
F) Amplifier (out)
G) Keyboard CV (out)
H) Keyboard gate (out)
I) Sawtooth frequency (in)
J) Square frequency (in)
K) Filter (in)
L) Filter frequency (in)
M) Amplifier (in)
N) Amplifier CV (in)
O) Envelope gate (in)
P) Main output stage (in)

As before, make yourself a label and Blu Tack it to the breadboard.

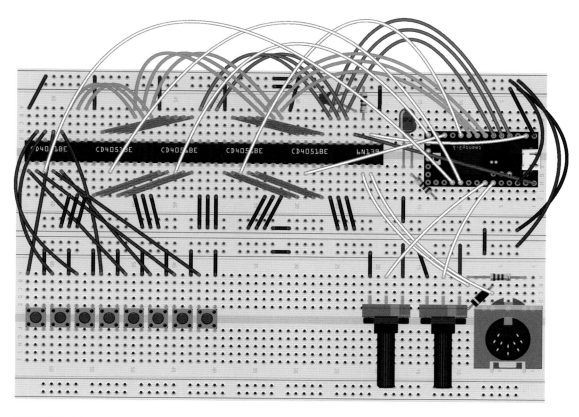

Figure 1 ◈
A diagram of the full synth (audio board and patch point wiring omitted for clarity) – note the diode, resistors, and capacitor required for the MIDI input

SPOT THE DIFFERENCE

Download the code from **hsmag.cc/issue17** and have a look at it – there are quite a few differences from part one. Firstly, the audio connection code (generated by the online Teensy audio design tool) has been moved to a separate file. This is because there are a lot more virtual connections this time, so keeping them in their own file makes the main sketch look a lot tidier. The code that handles polyphonic note data from the keyboard has also

been moved to separate files. Another new element is the MIDI library, which is included and initialised at the top of the code.

The next change is that the `inputMixers` array is now a much more complicated, multidimensional array. Instead of being a simple list of references to four modules, it now contains two separate arrays, which we need because we are creating a polyphonic (multi-note) synth with two copies of every module.

The other most significant difference is that the main **for** loop is now more complex. Previously it was a nested **for** loop with two levels, which was fine because we only had one chip sending data and one chip reading it, but our new circuit →

Figure 2 ◈
There are 16 'patch points' which connect to each other, creating the signal chain – run jumper wires from here to the second breadboard

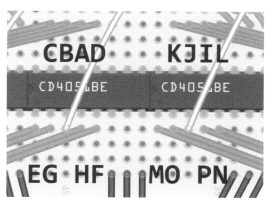

HOW DOES POLYPHONY WORK?

A lot of classic synths, and the vast majority of modern modular synths, only play one note at a time. When designing a synth that plays multiple notes at once, you have to consider what the maximum number of notes playable will be, and which notes should be silenced if you go beyond this maximum.

In this synth, we have created two copies of every virtual module in the code, giving two-note polyphony. Try holding down three or more notes and see what happens. If you want to change the current behaviour, for instance to prioritise the highest note, you can edit the **KeyboardHandler** class files.

This synth's polyphony has been kept at two notes to make the code easier to understand, but you should be able to increase it to four notes or even more by tweaking the code, without changing anything in the circuit.

```
Patch 1: "sci-fi" test
A (LFO) ─────────➤ J (square CV)
C (square) ─────➤ P (main output)

Patch 2: keyboard test
G (key CV) ─────➤ I (saw CV)
H (key gate) ──➤ O (envelope gate)
E (envelope) ──➤ N (amplifier CV)
B (saw) ────────➤ M (amplifier)
F (amplifier) ─➤ P (main output)

Patch 3: throbbing filter
G (key CV) ─────➤ J (square CV)
A (LFO) ────────➤ O (envelope gate)
E (envelope) ──➤ L (filter CV)
C (square) ─────➤ K (filter)
D (filter) ────➤ P (main output)
```

Above ⬈
Wondering what to do with your new synth? Here are some things to try

Right ◈
We've kept the patch points in order for this design (outputs on left, inputs on right), but you can easily rearrange them into a more convenient order

WHAT IS MIDI?

MIDI stands for 'musical instrument digital interface', and is a system whereby one instrument can control another via a special cable. The MIDI standard can deal with all sorts of musical information, such as tempo, pitch bend, and sustain pedal, but for this synth we're just going to implement the basic 'note on' and 'note off' commands.

necessitates a four-level loop. The principle is the same, but we are having to alternate which chips are active at a given time, hence the extra levels.

Inside the `for` loop, we also check for incoming MIDI data, pass it to the `KeyboardHandler` class so that polyphony is handled correctly, then convert the notes to virtual CV and gate signals so that they can be used for patching.

PLUG IN, BABY

Upload the code to the Teensy. If all has gone well, you should now have a synth that is very similar to part one, but with 16 patch points. Try some simple patches, such as the square wave going straight to the output – this should produce a simple tone. Now, referring to patch diagram, recreate patch 1

> **It makes sense to build the circuit step by step,** so we can check for errors

using jumper wires, and adjust the right-hand potentiometer – if it sounds like sci-fi effect, it's probably working. If not, check your connections.

Next, add and connect the final 4051 chip, plus the eight buttons that constitute our miniature keyboard. You should now be able to use patch point G (keyboard CV) to control the frequency of your oscillators, and patch point H (keyboard gate) to control the amplifier or envelope generator – try recreating patch 2 to see the keyboard in action. The sketch will allow the breadboard keyboard to function until a MIDI signal is detected, at which point the breadboard keyboard will be disabled.

Finally, add the MIDI input components. Because a MIDI input allows us to connect to another device, we use an optoisolator, which turns the incoming data into a series of pulses of light, then back into a digital signal again. If you would like more detail or ideas for troubleshooting, go to **hsmag.cc/vTjPpc** – the MIDI circuit for this synth was based on this design.

If everything seems to be working, congratulations! You have built a semi-modular polyphonic digital synthesizer, and you're ready to make the world a more musically interesting place. ▫

WHAT TO DO NEXT

There are loads of things you could do next with this synth. You could add code to make it recognise more MIDI commands, allowing MIDI control of the filter and envelope. You could change what the patch points do – perhaps you would like a white noise generator instead of a second oscillator? If so, have a look at **hsmag.cc/WzjFUw** – there are lots of virtual synth building blocks for the Teensy detailed there.

Perhaps the most satisfying next step, though, would be to upgrade this design from a pretty mess of breadboard wiring to a more permanent form using stripboard. You could keep using jumper wires for patching, while soldering everything else in place, and make a sturdy enclosure from wood, metal, or 3D-printed plastic.

TUTORIAL ▬▬▬▬▬▬

Build an ISS count-down timer

Know when the International Space Station is passing overhead with this timer

Ben Everard

🐦 @ben_everard

Ben loves cutting stuff, any stuff. There's no longer a shelf to store these tools on (it's now two shelves), and the door's in danger.

The International Space Station (ISS) is one of the greatest achievements of mankind. Ever. It's certainly one of the most expensive with an estimated cost of around $150 billion, and it's continuously housed people in space since November 2000. One of the most impressive things about it is that it can be seen from Earth. Every 90 minutes, it completes an orbit of the Earth, but the Earth is rotating underneath it, so each orbit goes over a slightly different section of the Earth.

Depending on where you are, and the particular week and month it is, you will probably find that you can see the ISS a few times a month, but when? We'll build a notifier that lets you know when the ISS will next be overhead.

Fortunately, the hard work of working out when the ISS will be overhead has been done for us, and there's a web API we can call to get the information. For this to work, you need to know your latitude and longitude. For example, in Bristol we can use the following URL with the relevant 'lat' and 'lon' values:

api.open-notify.org/iss-pass. json?lat=51.45&lon=2.89

You can enter this in a browser, and you'll get a response something like this:

```
{
  "message": "success",
  "request": { "altitude": 100,  "datetime":
1552572620,  "latitude": 51.45,  "longitude":
2.89, "passes": 5  },
  "response": [
    {
      "duration": 551,
      "risetime": 1552601375
    }
...
```

This encoding is called JavaScript Object Notation (JSON), and it's pretty common on web APIs, as it allows structured data to be passed via text. The important information for us is in the response section. This gives us a list of the next five times the ISS passes over us, with a duration of the pass (which we won't use), and the time it'll first be visible (which we will).

You can get this from any internet-connected device, and you could build this notifier on a microcontroller, but we'll be using a Raspberry Pi. Any of the WiFi-enabled models should work, but we'll use a 3B+.

We'll be using Python 3 for the brains of our project. The very basic code that grabs the next rise time is:

```
import urllib.request
import json


lat=51.45
long=2.89

ISS_url = "http://api.open-notify.org/iss-pass.jso
```

Below ◆
You can use any code editor you like. We've used Mu, but most programmers' editors work well with Python

```
untitled  ×  iss.py ×
1 import urllib.request
2 import json
3 import time
4 import board
5 import neopixel
6 from math import floor
7
8
9 pixels = neopixel.NeoPixel(board.D18, 6)
10
11
12 lat=51.45
13 long=2.89
14
15 colour_days = (50,0,0)
16 colour_hours = (0,50,0)
17 colour_tenmins = (0,0,50)
18 colour_mins = (15,15,15)
19 colour_now = (50,50,50)
20
21 ISS_url = "http://api.open-notify.org/iss-pass.json?lat="+str(lat)+"&lon="+str(long)
22
23 wait_time = 1
```

Python 🐍

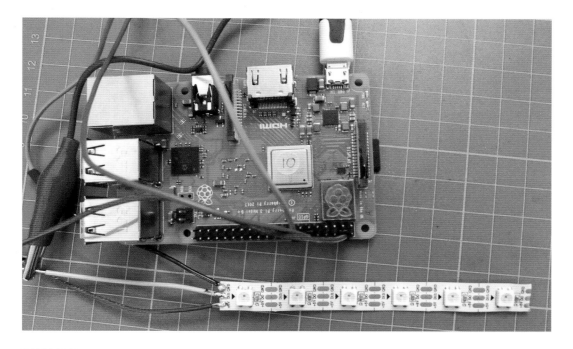

```
n?lat="+str(lat)+"&lon="+str(long)

with urllib.request.urlopen(ISS_url) as url:
    data = json.loads(url.read().decode())
    print(data)
    print(data['response'][0]['risetime'])
```

As you can see, this uses the JSON module to parse the response from the API and convert it into a Python data structure that is a combination of lists and dictionaries. We can then grab the bit of data we want out of this data structure.

GETTING PHYSICAL

The next step is to create the data display. You can obviously use whatever data output device you want, but we're going to go with a row of six NeoPixels. The display will be:

- If there's more than a day until the next time the ISS passes overhead, all will be red
- If there are between one and 24 hours, the LEDs will show a binary count-down of hours in green
- If there are between ten and 60 minutes, the LEDs will show a binary count-down of ten-minute periods in blue
- If there are between one and ten minutes, the LEDs will show a count-down of minutes in white
- If there's less than one minute to go, all the LEDs will glow bright white

There's obviously quite a lot of binary count-downs in that list, so we need a way of converting a time period into data that we can send to our NeoPixels.

 Java Script Object Notation (JSON) is pretty common on web APIs as it allows structured data to be passed via text

The NeoPixel module takes a list of tuples, where each tuple contains three digits: one for red, green, and blue. We can generate this list with the following function:

```
def countdown_leds(seconds, period, colour):
    led_array = []
    binary_string = "{0:b}".format(floor(seconds/
period)).zfill(6)
    for digit in binary_string:
        if digit == '1':
            led_array.append(colour)
        else:
            led_array.append((0,0,0))
    return led_array
```

The key part of this function is the line:

```
binary_string = "{0:b}".format(floor(seconds/
period)).zfill(6)
```

This uses a slightly obscure part of the **format** string function to convert the number into a binary string. The **format** function takes some data and puts it in a placeholder in a string. The standard placeholders are {0} .. {n}. In our case, our string is only a placeholder, but this doesn't have to be the case, →

YOU'LL NEED

◈ **Raspberry Pi**

◈ **Strip of 6 NeoPixels**

◈ **Connecting wire**

it could have been: "the binary digit is {0:b}". The placeholders can also take specifiers which, in our case, is a b which specifies that we want to insert the number as a binary string. The last part uses **zfill** to fill out the left-hand part of the string with zeroes until it's six digits long.

The main part of our code needs a few bits to work, including some modules and variable definitions:

```
import urllib.request
import json
import time
from math import floor

lat=51.45
long=2.89

colour_days = (50,0,0)
colour_hours = (0,50,0)
colour_tenmins = (0,0,50)
colour_mins = (15,15,15)
colour_now = (50,50,50)

ISS_url = "http://api.open-notify.org/iss-pass.jso
n?lat="+str(lat)+"&lon="+str(long)

wait_time = 1
```

This pulls in a few modules, defines the URL (you'll need to change the latitude and longitude for your location), and defines the colours we'll use. The only unusual bit there is the **wait_time** variable. This holds the number of seconds our code will wait between calls to the API. We keep calling the API to get the latest information, but we don't want to hammer the API, so we pause for a given number of seconds between calls. The API is rate-limited to one request a second, so we shouldn't exceed that, but really,

VOLTAGE PROBLEMS

Technically, NeoPixels require the data signal to be at least 0.7 times the drive voltage. We'll be driving them off the 5 V pin (as the 3 V pin isn't enough to drive them). This means that we should have a data signal of at least 3.5 V. The GPIO pins on the Raspberry Pi are only 3.3 V. Usually, we find this doesn't cause a problem, but if your NeoPixels are behaving badly, you may need to do something to either decrease the drive voltage, or increase the data voltage. There are some tips for doing this here: **hsmag.cc/eYvUWK**.

we don't need to even be close to that. We'll use a sliding wait period, where the longer until the next predicted passover, the longer we'll wait.

The code for the main loop is then:

```
while True:
    time.sleep(wait_time)
    with urllib.request.urlopen(ISS_url) as url:
        data = json.loads(url.read().decode())
        next_vis=data['response'][0]['risetime']
        seconds_to_next_vis=data['response'][0]
['risetime'] - time.time()
        if seconds_to_next_vis > 86400:
            neopixel_array = [colour_days]*6
            wait_time = 600
        elif seconds_to_next_vis > 3600:
            neopixel_array = countdown_
leds(seconds_to_next_vis, 3600, colour_hours)
            wait_time = 60
        elif seconds_to_next_vis > 600:
            neopixel_array = countdown_
leds(seconds_to_next_vis, 600, colour_tenmins)
            wait_time = 10
        elif seconds_to_next_vis > 60:
            neopixel_array = countdown_
leds(seconds_to_next_vis, 60, colour_mins)
            wait_time = 2
        else:
            neopixel_array = [colour_now] * 6
            wait_time = 2
    #write LEDs to neopixels
    print(neopixel_array)
```

As you can see, this has one big if-elseif-else statement that calculates the correct lighting for the particular time. If you bring this all together, you'll get the code for turning the NeoPixels on and off. Now, let's look at the hardware.

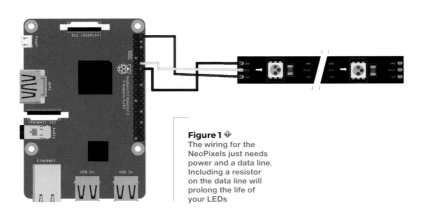

Figure 1 ◈
The wiring for the NeoPixels just needs power and a data line. Including a resistor on the data line will prolong the life of your LEDs

> **You can cut the LED strip to any length;** just cut through the exposed pads

CONNECT THE LEDS

First, we'll need six NeoPixels. These come in a wide range of forms, but we'll go with a common LED strip cut so there's six. You can cut the LED strip to any length; just cut through the exposed solderable pads, making sure that there's enough to solder onto on each side. Make sure you solder to the correct end of the strip: there are data-in and data-out sides. Solder to the data-in side (there may be an arrow indicating the direction of data flow). You should have a power wire, a ground wire, and a data wire. It's good practice to include a 330 Ω resistor on the data wire before the first pixel, but this isn't necessary for it to work (it will mean the pixel lasts longer).

The 5 V wire can be connected to a 5 V pin on the Raspberry Pi (see the Voltage Problems box on previous page), and the ground to a ground pin. The data connection should go to pin 18 (see **Figure 1**).

There's a bit of software you need to install to make it all work. Type the following into a terminal session on the Raspberry Pi:

```
sudo apt install python3-pip
sudo pip3 install rpi_ws281x adafruit-
circuitpython-neopixel
```

You can check that this is all working by opening a Python 3 prompt with root permissions. Root is needed to control the NeoPixels; enter `sudo python3`. Now, enter the following code:

```
>>> import board
>>> import neopixel
>>> pixels = neopixel.NeoPixel(board.D18, 6)
>>> pixels[0]=(0,10,0)
```

This should turn the first NeoPixel on green. You can use this to turn on any pattern you like. Once you've finished playing, let's get back to our main code.

You need to include the lines to set up the NeoPixels directly under the other `import` blocks:

```
import board
import neopixel

pixels = neopixel.NeoPixel(board.D18, 6)
```

```
47
25 def countdown_leds(seconds, period, colour):
26     led_array = []
27     binary_string = "{0:b}".format(floor(seconds/period)).zfill(6)
28     for digit in binary_string:
29         if digit == '1':
30             led_array.append(colour)
31         else:
32             led_array.append((0,0,0))
33     return led_array
34
35 while True:
36     time.sleep(wait_time)
37     with urllib.request.urlopen(ISS_url) as url:
38         data = json.loads(url.read().decode())
```

Now, you just need to write the values from the `neopixel_array` list to the `pixels` list. Add the following directly under `print(neopixel_array)` at the end of the main loop.

```
print(neopixel_array)
for i in range(0,6):
    pixels[i] = neopixel_array[i]
```

That's all there is to the main code. You can run this by entering:

```
sudo python3 iss.py
```

You'll need to change **iss.py** to whatever you've called the file. Running like this works, but you'd need to re-run this command every time you turned on the Pi. Obviously, not ideal for something that's designed to be embedded. There are a few ways of configuring your Pi to run a script every time you turn it on, but one of the easiest is to add a line to the **/etc/rc.local** file. You'll need to edit this as root, so enter the following in a terminal:

```
sudo nano /etc/rc.local
```

This opens the nano text editor and lets you edit the file. Immediately above the line `exit 0`, add:

```
python3 /home/pi/iss.py &
```

Here, the code is saved as **iss.py** in the Pi user's home directory, but you can change this if you prefer. The **&** at the end tells the script to continue running and run our command in the background – essential as **rc.local** will block the boot process if it doesn't finish. Hit **CTRL+X** to save and exit.

That in place, you can restart your Pi, and it should run the script automatically. All you have to decide is how to mount and show off your ISS notifier! □

Above ◈
One function takes care of the conversion from seconds to a binary clock

PLA spring motor

 By **Greg Zumwalt** ⟡ hsmag.cc/ldfD0o

" **A**fter spending years designing and programming video games for Tandy, Atari, Electronic Arts, and many others, plus even more years designing flight control systems, I retired, purchased a 3D printer, downloaded a CAD program, and started designing models for 3D printing.

"In my early 3D printing days, most of my design and 3D printing time was spent trying to determine what I could and could not 3D-print in PLA on a 3D printer. In early 2014, I began testing whether a 3D printer could print a PLA spring motor, and the PLA Spring Motor Demonstrator was the result.

"I was impressed enough with the PLA Spring Motor Demonstrator test to attempt to design, 3D-print, and assemble a variety of wind-up vehicles, including a wind-up car, wind-up bunnies, wind-up helicopters, and more.

"I recently decided to revisit the original PLA Spring Motor Demonstrator design, using Autodesk Fusion 360, and the PLA Spring Motor Demonstrator II is the result.

"The design incorporates the use of Fusion 360 'parameters', allowing me to alter sizes and clearances between the gears, axles, and axle mounting holes, with ease. As a result of this update, it will be much easier to design and print PLA Spring Motor-based models." □

Right ⊿
Greg's done some fantastic work with springs, motors, and 3D printing:
hsmag.cc/jxsiFI

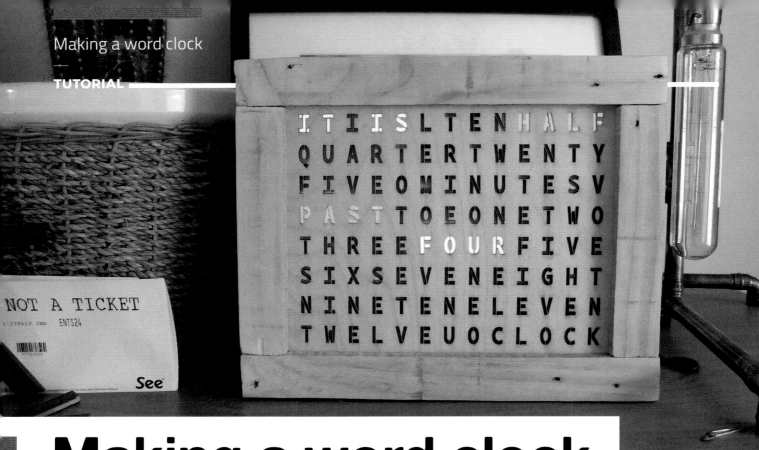

Making a word clock

Build your own attractive timepiece

Ben Everard

🐦 @ben_everard

Ben loves cutting stuff, any stuff. There's no longer a shelf to store these tools on (it's now two shelves), and the door's in danger.

O K, settle in. This project turned out to be a bit more complex than expected. Actually, complex isn't quite the right word. There's nothing in here that's fundamentally hard, but it did test our skills in quite a few different areas of making, and each area posed its own little challenges that needed to be overcome. We'll guide you through it as best as we can.

In this project, we used quite a wide range of equipment and parts. These represent the tools and parts we had available, rather than a canonical set of things you actually need. There's no 'right' way of doing this, and you can find alternatives to almost everything we've used if you need to.

The basic way a word clock works is that it shines a light through letters spelling out the words to say the time. The heart of our clock, then, is these letters and the LEDs to make the light. We used laser-cut 3 mm plywood for our clock face, but other people have had success using printed acetate sheets (the sort used in overhead projectors that older – but not too old – readers will remember from their school days). Thinner laser-cut sheets would also work, but we'd recommend going no thicker than 3 mm as this will reduce your viewing angle.

You can grab our design from **hsmag.cc/issue20**, but it's fairly easy to create your own (or modify ours if you'd prefer). The crucial point for the lettering is that we need to use a stencil font – this ensures that there's a connection to any isolated parts of a letter (such as the middle of the letter O), so they don't fall out when laser-cut. It makes layout easier if the font is monospaced – we used BP Mono Stencil (**hsmag.cc/BPMonoStencil**).

The LEDs must be held in the appropriate place behind the letters. There are two approaches that you can take here – you can design your letters so that they line up with off-the-shelf LED strips, or you can use strings of LEDs to line up with whatever spacing you use for the letters. We opted for the latter, but the former would make a more

Left ◈
The LED string
inserted. We had
to join three strings
together to get 104
LEDs for the clock.
In hindsight, only
100 are needed, as
some letters are
never lit

YOU'LL NEED

◈ **WiFi-enabled microcontroller (such as the MKR1000)**

◈ **String of 104 NeoPixels**

◈ **1 A diode**

◈ **9 mm plywood**

◈ **3 mm laserply**

◈ **Wood for frame**

◈ **Laser cutter**

◈ **Modelling foam**

straightforward build if you're less fussy about the size of the clock.

We then need a way of holding the LEDs in place behind the letters. There are a few parts to this – first, you need a way of holding the LEDs in place far enough behind the letters so that they illuminate them evenly; then you need a way of minimising the amount of 'bleed', where lighting one letter illuminates the letters either side of it; finally, you need something to diffuse the light.

Our setup used plywood with 7 mm holes drilled into it. This is just large enough for surface-mounted 5050 LEDs to be pushed in place and held with a drop of superglue. These shone through the holes in the plywood and into a square honeycomb made of modelling foam hot-glued together. Finally, it hit a double-layer of diffusion fabric before shining through the laser-cut face. All we needed was a frame to hold it in place. We made this from 4×1 inch reclaimed wood with routed grooves to hold the face and plywood LED panel in place.

Let's take a closer look at this process before diving into the microcontroller brains.

THE BUILD

First, you'll need to laser-cut your clock face – that's the easy bit of woodwork. Now on to the manual part…

As mentioned, we started building our frame with reclaimed 4×1 inch wood that cost just £1 from our local wood recycling project. We sanded this down to give it a smooth finish, but it lacks the hard corners of planed wood. There are also holes from old nails which combine with the rustic

joining technique to give the look we wanted for our clock.

If you're an experienced woodworker, you may choose a more elegant method for making the frame, but as we're not, we'll keep it simple. We've used butt joints in the corners which are held together with two screws each. First, we routed two grooves in one side of the wood – one to hold the 3 mm face, and one to hold the 9 mm

 It hits a double-layer of diffusion fabric before shining through the laser-cut face

plywood LED panel. 9 mm ply is overkill for such a frame, but we happened to have some spare from a previous project – you could easily get by with 3 mm or 6 mm plywood, and MDF would work just as well. We routed these grooves 3 mm deep into the wood. →

WIRING

The simplest wiring for the clock is to connect the 5 V and GND pins and one data point (we used pin 6) from the microcontroller to the 5 V, GND, and data input pins on the first LED. The LED chain will then propagate power and data along the strip. However, there are a few problems with this.

Firstly, this results in an out-of-spec power situation which you might, or might not get away with (see 'Power problems' box on page 109). Secondly, the jitter on the power line may cause problems – putting a capacitor between the 5 V and GND lines can smooth that. Thirdly, you should include a 470 Ω resistor between the Arduino pin and the data-in line. You might get away without this, but it will prevent any problems with too much current being drawn.

If you've got a plunge router bit, you might choose to do this groove-cutting later, and not rout all the way to the edge of each section of frame as this will give a better finish.

We then needed to cut the wood into four appropriate length sections. You need two for the top and bottom that are:

length = width of face + (2 × width of frame wood) − (2 × depth of groove)

And two for the side that are:

length = height of face − (2 × depth of groove)

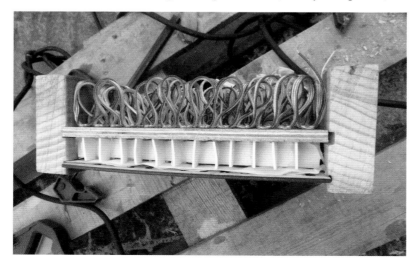

You should now be able to hold them all by hand and everything should fit together (don't screw or glue them together yet). If they don't fit, you'll need to make adjustments before moving on. This might entail routing the grooves a little deeper, or trimming down the wooden frame.

INNARDS

The quickest way to mark up the plywood LED holder is by eye. It needs to be the same size as the clock face, and you can pencil-mark the spots for the LEDs very quickly without the need for a tape measure (though measure and mark properly if you'd prefer).

As previously mentioned, we drilled these out with a 7 mm drill bit. The diagonal on a 5050 surface-mount part is just larger than 7 mm, so it's a tight fit. We used strings of WS2812 LEDs (often known as NeoPixels). Each LED is on a small, circular PCB. We applied a drop of superglue to the edge of each LED, then pushed it into the hole in the circuit board. They take a bit of force to get in, but be careful, as we pushed too hard on one and dislodged a resistor (if you do this, just cut out the LED in question and join the wires with solder).

TESTING

Your build will almost certainly be slightly different to ours, so rather than just following along by rote and hoping that the results are the same, now's a good

challenging than we anticipated, but with the right technique it's not too hard.

First, anchor one end of one long strip to the frame and wait for the glue to harden. Then put a 'U' of glue in where you want one of the separators to go, and then slot the separator into this glue (don't try to hold it in place while you put the glue in). With practice, you can do several of these 'Us' of glue at a time (we found four or five was a good number), then insert all the separators in one go. Before you finish one row, anchor the next long row strip to the frame, as this gives it time for the glue to harden before starting that row.

point to pause and check that everything's working as you'd like.

Connect the microcontroller up to the NeoPixels (we used crocodile clips, but you can solder it up if you don't have these). See the 'Wiring' instructions box on page 105.

We used the test code from the Adafruit NeoPixel Überguide to make sure everything was working properly (**hsmag.cc/ArduinoLibraryUse**). Bear in mind that lighting up all the pixels at once will take quite a bit of current, so you will want to either use an external power supply or dial down the brightness (we tested ours with a colour of (10,10,10) and this worked with the on-board regulator on the MKR1000).

With this and a mess of wires in place, and everything working, let's move on with the assembly. Screw together three sides (one long side and two short sides) of the frame. To ensure that it is in the right place, it's a good idea to use an F-clamp to hold it together with the clock face and the plywood panel in place while drilling and screwing.

Leave one F-clamp in place, holding the two ends of the wood on the exposed side together while finishing the internal assembly.

We used 1 mm-thick white modelling foam for the square honeycomb inside the frame. You may want to consider laser-cutting this out using something like the tray insert pattern from **hsmag.cc/TrayInsert**; however, we didn't. We cut long strips the width of the frame and the height of the gap between the plywood and the back of the face, and small 'separator' strips to split it up vertically. Gluing this together was a bit more

> **" Experiment with what you have** to see what creates the aesthetic that you want **"**

DIFFUSION
The final thing to add before assembly is diffusion. This can be anything that's translucent and thin enough to fit in the space. We used photographer's diffusion fabric (essentially a thin, white nylon material), and we found that we needed two layers of this to get the look we wanted, but it's not standard fabric, so experiment with what you have to see what creates the aesthetic that you want.

We cut this to size and placed it over the square honeycomb. A few dabs of hot glue on the corners held it all in place (and this won't be visible once it's fully assembled). →

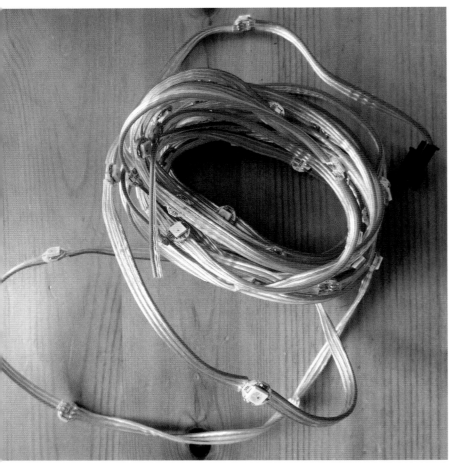

Above ◈
NeoPixel strings are easier to mount in custom spacings than NeoPixel strips, but still have the advantage of not having to solder every LED

```
<?xml version="1.0" encoding="ISO-8859-1"?>
<result>
<status>OK</status>
<message/>
<formatted>2019-05-30 14:52:07</formatted>
</result>
```

There are two parts to getting and processing this in Arduino. First, we have to download this XML, and then we have to extract the time from it. The method of connecting to WiFi differs a little depending on what hardware you're using. We used the WiFi101 library, but if you're using different hardware (such as an ESP8266) you might need to do it slightly differently. Take a look at your board's example WiFi sketches for details.

Once connected, we have a client object linked to the **api.timezonedb.com** server (see the full code for more info on this). We can then extract the appropriate line in the response with the following:

```
    client.println("GET /v2/get-time-zone?key=YOUR
KEY&format=xml&fields=formatted&by=zone&zone=Euro
pe/London HTTP/1.1");
    client.println("Host: api.timezonedb.com");
    client.println("Connection: close");
    client.println();
    }
delay(10000);
payload = "";
    Serial.println("stand by for data");
    while (client.available()) {
    char c = client.read();
    Serial.write(c);
    if (c == '\n') {
    payload = "";
    }
    payload += c;
    if(payload.endsWith("</result>")) {
    parse_response();
    }
```

This reads the HTTP response character by character and builds up a string called **payload**. If it reaches a newline character, it empties **payload** as we only want one line. If it reaches the string **</result>**, it knows that it's got the data it needs, so it called the function **parse_response**.

The key parts of this function are as follows:

```
    int colon = payload.indexOf(':');
// Set the first colon in time as reference point
    nowday = payload.substring(colon - 5,
```

Once you're happy with the amount of diffusion, you can attach the final side of the frame, and that's the hardware setup complete. Now let's take a look at the software.

The full code for this is available from **hsmag.cc/ClockCode**, but let's take a look at the most pertinent bits.

Obviously, our clock needs to know what the time is. We could have used a real-time clock, but this would still necessitate setting the clock time manually and adjusting the time for daylight savings. Instead, we decided to grab the time from the internet – specifically, **timezonedb.com**.

You'll need to register for a free API key, but we'll be staying well within the limits of free use. Once you've got that, you can grab the current time in a particular location by pointing your web browser to **api.timezonedb.com/v2/get-time-zone?key= KEYHERE&format=xml&fields=formatted&by= zone&zone=Europe/London**.

You'll need to replace **KEYHERE** with your key, and if you're not in the UK you'll need to update the zone to your location. The result comes back in XML, and should be something like:

```
colon - 3);
    d = nowday.toInt();
    nowmonth = payload.substring(colon - 8,
colon - 6);
    mo = nowmonth.toInt();
    nowyear = payload.substring(colon - 13,
colon - 9);
    y = nowyear.toInt();
    nowhour = payload.substring(colon - 2, colon);
    h = nowhour.toInt();
    nowmin = payload.substring(colon + 1, c
olon + 3);
    mi = nowmin.toInt();
    nowsec = payload.substring(colon + 4,
colon + 6);
    s = nowsec.toInt();
```

Since the time and date is in a specific format, we can locate the particular part we want relative to the first colon. This code pulls the string apart and converts the relevant segments into integer values for the hours, minutes, and seconds. It also extracts the date, but we don't use that. We adapted this code from Arduino forum user Aggertroll – thanks Aggertroll!

This reads the HTTP response character by character and builds up a string called 'payload'

Now that we've got the time, we need a way of displaying it on the NeoPixel strip. This is done by first creating a series of arrays that hold the locations for the pixels in different words, such as:

```
int itis[] = {8,9,11,12};
int five[] = {35,36,37,38};
int ten[] = {4,5,6};
```

We also created a function that turns the LEDs in one of these arrays a specific colour:

```
void lightup(int letters[], int letters_len, int
red, int green, int blue) {
    for(int i = 0; i<letters_len; i++) {
    strip.setPixelColor(letters[i], red, green,
blue);
    }
    strip.show();
    }
```

POWER PROBLEMS

Once we wired up our clock, we found that it frequently glitched out and flashed strange colours. After unsoldering all the connections and rewiring it all up, we realised that the problem wasn't a cold joint, or even code problems, but a voltage mismatch.

We powered the LEDs from the 5 V pin on the microcontroller (we can keep the LED numbers and brightnesses sufficiently low to allow this to work); however, the data pins on the MKR1000 are 3.3 V. The input to the LEDs should be (according to their datasheet) at least 0.7 times the power voltage (3.5 V), so we're going out-of-spec by powering it at 3.3 V. Usually we can get away with this, but the particular LEDs we used proved to be particularly finickety about this.

There are two basic solutions to this – increase the input voltage or decrease the power voltage. We opted to do the latter by putting a diode with a forward voltage of 0.8 V on the power line. This diode has to be able to take the full current of the LEDs (we used a 1 A diode, which should give us plenty of leeway). Alternatively, you can use a level shifter (these are available in both module and IC form) to increase the voltage from the data signal to 5 V.

The final code for lighting up the correct time is as follows:

```
strip.fill();
lightup(itis, 4, 100,100,0);
int hour = h;
if (mi > 33) { hour+=1;}
 if (hour > 12) { hour -= 12;}
if (hour==1) { lightup(h_one, 3, hour_red, hour_
green, hour_blue); }
if (hour==2) { lightup(h_two, 3, hour_red, hour_
green, hour_blue); }
...
//past or to?
if (mi > 3 && mi < 34) { lightup(past, 4,0, 150,
0); }
if (mi > 33 && mi < 58) {lightup(to,2,0,150,0);}
if (mi > 57 || mi < 4) {lightup(oclock,6,50, 50,
100);}
// minutes
if (mi > 3 && mi < 8) {lightup(five, 4, mins_red,
mins_green, mins_blue); lightup(minutes, 7,mins_
red,mins_green, mins_blue);}
if (mi > 7 && mi < 14) {lightup(ten, 3, mins_red,
mins_green, mins_blue); lightup(minutes, 7,mins_
red,mins_green, mins_blue);}
...
```

The first line in this code blanks the whole strip, then the line `lightup(itis, 4, 100,100,0);` lights up the words 'it is'. We then have to find the first hour, bearing in mind that as soon as the minutes have gone past 34, it will switch to 'twenty-five to' the next hour. The code then ends with a series of **if** statements that find the correct letters. ◻

Make a Slim Jim antenna

Let's make a simple, and affordable, starter antenna for a SatNOGs ground station

Jo Hinchliffe

🐦 @concreted0g

Jo Hinchliffe is a contributor to the Libre Space Foundation, and is passionate about all things DIY space. He loves designing and scratch-building both model and high-power rockets, and releases the designs and components as open source. He also has a shed full of lathes, milling machines, and CNC kit.

Right ◈
The Slim Jim antenna, seen here coiled up with the rest of a simple SatNOGS ground station kit

YOU'LL NEED

◈ **2 metres of 300 Ω ladder feeder cable** (available quite cheaply in short lengths online)

◈ **50 Ω RG58 BNC cable**

◈ **Cutters**

◈ **Soldering iron**

◈ **Electrical tape**

Earlier this issue, we walked through setting up a SatNOGS ground station and, whilst many people buy an antenna to complement their station, it's possible to begin hunting satellites by making a simple 'Slim Jim'-type antenna for less than £15.

There are numerous tutorials and calculators online for these antennas, and they are all worth checking out. The dimensions for the antenna we will build are based on the figures from a calculator at **hsmag.cc/YnLbhD**, which I found via an excellent Essex Ham tutorial here: **hsmag.cc/muniTO**.

We are going to aim to make an antenna that is sized and tuned so that it should be optimised

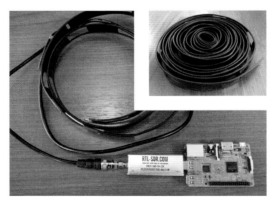

TUNE **IN**

There are many frequency bands that satellites operate on, and we have opted to build an antenna that suits part of the VHF (very high frequency) range. Another common band is UHF, which is even higher (ultra high frequency). Due to the actual physical length of the radio waves at these frequencies, they often get referred to as the two-metre band or the 70 cm band respectively. It's worth knowing these terms, as they get used a lot!

to receive signals around 145.800MHz. This is a common area of frequency for many, many satellites, and also the International Space Station periodically transmits different signals on this frequency. So, setting the frequency to 145.800 in the calculator on the M0UKD website above, we are given the collection of dimensions we need and the layout of the Slim Jim antenna. The first task is to cut a length of the ladder feeder wire so that it is a little over the total length required. The length required is 150.2 cm, so cut a length to around 156 cm.

We then strip and form one end, so that the wires meet, and solder them together, as shown in **Figure 1**. Then, we measure to try to bring the antenna to its total length from the joint we have just made, allowing a short section to be left over to form the loop at the other end. Cut and strip, and solder the other end in the same way as the

Above ◈
The m0ukd.com website has a calculator which will give the dimensions of a Slim Jim-type antenna for any target frequency you submit. Go to this calculator (hsmag.cc/YnLbhD) and input 145.800 to get all the desired information

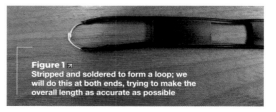

Figure 1 ↗
Stripped and soldered to form a loop; we will do this at both ends, trying to make the overall length as accurate as possible

first. Again, we are aiming for the overall length from the end of each soldered loop to be our target dimension of 150.2 cm, so try to be as accurate as possible!

Next, we need to cut a small gap in one side of the loop only. We measure up the distance given from the calculator, which was in this case 49.4 cm, from one end and then carefully cut out a 2.1 cm notch in one side of the antenna only! On our ladder

 We decided to add some support, keenly engineered from **a coffee stirrer and some tape**

wire, this notch happened to land on an unsupported section of the wire, so we decided to add some support, keenly engineered from a coffee stirrer and some tape!

GETTING IT CONNECTED

The next task is to clear/strip a small section for RG58 coaxial cable to be connected. If you have experience in attaching BNC connectors to coaxial cables, you might opt to make up your own cable here. But, in order to keep it simple, it is as cheap to buy an RG58 coaxial cable with a BNC connector on each end and cut it in half. Either way, we need to carefully clear a 3–5 mm section on both sides of the ladder feeder wire, so that we have just enough space to solder the RG58 cable on. This should be done so that the RG58 connections are as close to 4.9 cm from the end of the antenna as possible (between the gap we cut and its closest end). Solder the shield of the coaxial cable to the side of the antenna that has the gap we cut into, and solder the central core of the RG58 cable to the other side. It's helpful to perhaps tape the RG58 cable to the ladder cable prior to soldering, to help keep things steady, as shown in **Figure 2**.

These antennas are designed to be hung off things and so should be positioned vertically, and many people have made them more permanent and weatherproof by installing them into plastic tubes, etc. They are also very portable, in that they can be rolled up and chucked in a bag for mobile use. A rolled-up Slim Jim and a key-chain carabiner can make a very quick-to-deploy ad hoc antenna. We can see in **Figure 3** that, even though this antenna has been hung inside, it can still receive a signal from space! If you click through to the Audio tab at **network.**

READY **TO SEND**

If we were wanting to use this antenna to transmit, it would be worthwhile trying to check and assess its standing wave ratio, and perhaps adjust it to make it extremely efficient and accurate. People also fit chokes and filters, and perhaps also a low noise amplifier (LNA) to further enhance the performance. However, for our 'receive only' duties, it is less important, and we can hook it up to our SatNOGS ground station and give it a go!

Above ◈
The gap, cut and reinforced

Figure 2 ◈
Using some tape, to physically attach the RG58 cable, helps with soldering this step

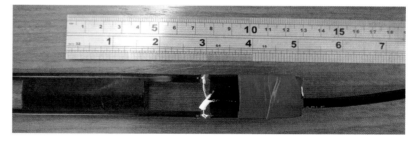

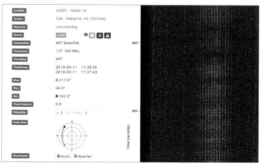

Figure 3 ◈
Here, we can see a successful observation using the simple SatNOGS test station we built earlier connected to the Slim Jim antenna

satnogs.org/observations/53010 and scroll to around three minutes, you will hear the Morse code telemetry beacon of a small satellite called XW-2F CAS_3F, captured on this antenna with the SatNOGS station built as part of the space cover feature in this issue. ☐

HackSpace
TECHNOLOGY IN YOUR HANDS

INSPIRATION

HACK | MAKE | BUILD | CREATE

Get inspiration for more projects with these guides to using household objects, plus expert making tutorials and introductions to classic techniques

Top Projects – Showcase

Arduinoflake

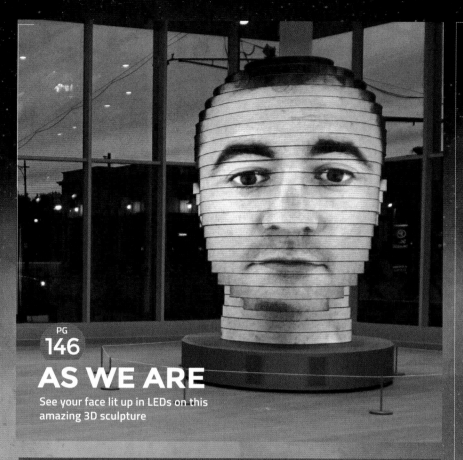

Top Projects – Showcase

Secure your builds with this crafty little fastener

Mayank Sharma

🐦 @geekybodhi

Mayank is a Padawan maker with an irrational fear of drills. He likes to replicate electronic builds, and gets a kick out of hacking everyday objects creatively.

or something that looks so basic, the clothes peg is a fairly recent invention. This all-important weapon in our battle with the wind made its debut only in the 19th century, which is a shocker, especially when you consider that hand grooming kits were in vogue in ancient Rome.

In its short period of existence, the clothes peg has essentially only had one revolutionary design upgrade. The first versions were a one-piece design, and they were commonly made by hand by the Romani in Europe. It's often claimed that sometime in the early 1800s, a patent for this design was given to someone named Jérémie Opdebec, about whom there is little to no information. These early one-piece clothes-pins were usually made of wood. Still in production these days, they are known as 'dolly pegs' and, in addition to wood, are also made from plastic.

Between 1852 and 1887, the US patent office issued almost 150 patents for clothes pegs. The more modern-styled clothes-pin came about in 1853. It was invented by David Smith, a serial American inventor,

and featured two separate pieces of wood and a spring. This design was tweaked in 1887 by Solon E Moore, who added the characteristic coiled fulcrum to join the two grooved pieces of wood at the centre of the clothes peg. Made from a single wire, the pegs were now stronger and cheaper to manufacture. However, the pegs had a limited lifetime, as they were prone to decay and rust. This changed with the invention of stainless steel. These days, clothes pegs come in a variety of shapes, sizes, and colours, but they all usually feature a spring between two wedge-shaped pieces of plastic, although you can still find some made from wood as well.

While made to secure wet clothes on a clothesline, the clothes peg has always been used for other purposes. Hollywood actress Diane Keaton has proudly admitted to using one to reshape her nose, and proponents of ear reflexology suggest using a clothes peg instead of finger pressure to one of the key pressure points. They are also famously used on movie sets to handle the hot diffusers on film lights. So, it really isn't a surprise that intrepid makers have clamped on to them.

IPAD PIANO

Project Maker
ADAM KUMPF

Project Link
hsmag.cc/KIEQRd

While writing code for musical instrument apps, Adam Kumpf was very conscious of the incomplete interaction between the users and the touchscreen: "I remember being constantly frustrated by the direction things were going with mobile tech design.** Next-generation tablets and smartphones seemed like a bland sea of flat, tactile-less interactions. I was spending my days writing code for musical instrument apps, and constantly felt the tension between all the capabilities of the device/software, and how flat everything felt (literally), because it was just fingers on a screen."

So Adam, who has a background in computer science and tangible interfaces from MIT, decided to make something that "resembled the interaction of a piano without changing the multitouch app on the tablet." He headed to the store and picked up a couple of inexpensive bits and pieces, and had a fully functioning keyboard by the next day. He used some aluminium foil and clothes pegs to trick the iPad into thinking it was his fingers playing the keyboard on the touchscreen. His Instructable has detailed explanations, and each step is illustrated to help you hack the clothes peg keyboard in a jiffy. □

"I REMEMBER BEING CONSTANTLY FRUSTRATED BY THE DIRECTION THINGS WERE GOING WITH MOBILE TECH DESIGN"

Below ◇
Adam has a couple of capacitive touchscreen hacks under his belt, and you can check out his current ones at Makefast Workshop

PEG HANDS

*I*t was when he was desoldering one of the tiny resistors on the ESP-12 chip that Suwardi Thio felt he could use an extra pair of hands to handle all the equipment. So, he decided to adapt his DIY Tool Organiser project, along with his ingenious multifunction twister wires, to create his own pair of helping hands. The wires are attached to a piece of board that Suwardi suggests should be at least 2 cm thick to better support the artificial limbs. You can drill holes in any pattern on the board, but he spaced the holes 5 cm apart from each other. To create the twister wires, he runs different lengths of the standard 2 mm electrical wire through small PVC hoses. One end of the wire is connected to the board, while the other is attached to the clothes peg via its spring hole. The exact process of connecting the wire to the board and the peg is nicely explained and illustrated in his Instructable. He uses wooden clothes pegs instead of alligator clips, since they are made of softwood and won't scratch the delicate electronics. □

Project Maker
SUWARDI THIO

Project Link
hsmag.cc/YUbwTF

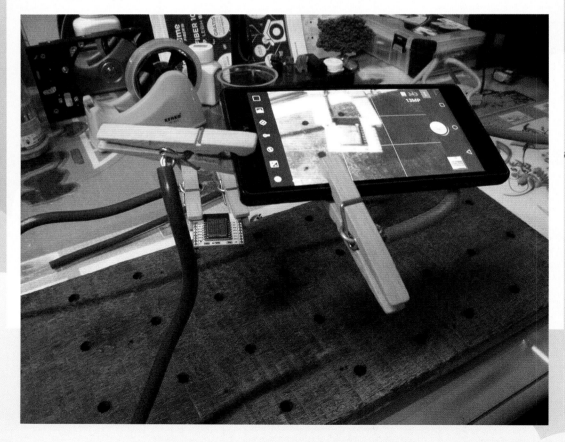

Left ◆
Suwardi's pair of peg hands are very sturdy and, besides handling electronics, he uses them for smartphone photography

CLOTHES PEG LAMPSHADE

J im is an old-school maker, and refuses to use modern-day conveniences like laser cutters, laser printing, CNC routers, and such: "My view is that the 'feeling' or individuality is lost from a piece of art or whatever you call it, using standardised machines." This is why in order to create the wire frame for his clothes peg lampshade, Jim shaped a galvanised steel wire into seven concentric rings by hand. To close the loop, he heated the fastening point where the wires meet and applied the solder. The seven concentric rings were connected via four vertical strips of wire using the same soldering mechanism. With the wire frame in place, he simply started clipping the clothes pegs to the ring at the bottom and worked his way up. Jim suggests placing the pegs carefully to ensure the ones in each subsequent row overlap the ones underneath it. ☐

Project Maker
JIM NEIL BERG

Project Link
hsmag.cc/sAviPN

Left ◆
Make sure the rings aren't too far from each other, or else the pegs will loosely dangle on the frame

SIMPLE DOOR ALARM

Z hyar is a 16-year-old maker from the Kurdistan region of Iraq, and hosts the Kurdish Inventor YouTube channel. One of his projects involves creating a simple door alarm using a clothes peg with copper strips attached to both its inner lips. The copper strips have small pieces of wires attached to them. Both are connected to a 9V battery, with one running via a 1kΩ resistor, an LED, and a buzzer. The whole contraption is mounted on a Styrofoam board and hot-glued to the side of a door. He wraps a piece of wire on the door handle, and connects the other end to a piece of card that's placed between the two copper strips on the clothes peg. When the door handle is turned, the piece of card is dislodged, and current flows between the two lips of the clothes peg, arming the buzzer and activating the LED. ☐

Project Maker
STUDIO STYLENRICH

Project Link
hsmag.cc/Oxweyh

Left ◆
The YouTube channel (hsmag.cc/arvnOO) has loads of electronics-related and other DIY projects

PAPER PLATES

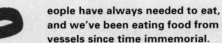

IMPROVISER'S TOOLBOX

You are not supposed to play with your food, but no one said anything about the utensils

Mayank Sharma

🐦 @geekybodhi

Mayank is a Padawan maker with an irrational fear of drills. He likes to replicate electronic builds, and gets a kick out of hacking everyday objects creatively.

People have always needed to eat, and we've been eating food from vessels since time immemorial. But plates, as we know them today, were invented in European potteries sometime shortly after 1708. And, even though paper has been around for thousands of years, paper plates are a 20th century invention. The idea of creating plates out of paper came to American inventor Martin Keyes, when he observed workers at a veneer plant in New York eating lunch on thin waste pieces of maple veneer. Keyes has often credited the ingenuity of these workers as the inspiration behind the idea for developing plates using moulded pulp.

Keyes spent the next couple of years perfecting a machine that would mash wood pulp and mould it into paper plates. He had to then fight a court battle before he could patent the machine, since someone had stolen his idea and created a similar machine. As an aside, Keyes' courtroom battle underlines the importance of documentation, since it was his daily log, as he laboured to perfect his machine, that convinced the court of his claims.

Armed with the patent, Keyes set up a small mill and started producing paper plates in 1904. His first paper plate moulding machine was capable of

manufacturing 50,000 paper plates per day. It was, however, an unfortunate event a couple of years later that helped scale his business. The large-scale rescue efforts in the aftermath of the San Francisco earthquake in 1906 created a huge demand for paper plates, and led to Keyes setting up multiple manufacturing units.

These days, paper plates adorn kitchen cupboards around the world. They might not be as valuable as your prized china, but their convenience is unparalleled. Over the years, paper plates have become an umbrella term that encompasses all kinds of single-use disposable plates. Generally speaking, paper plates aren't very expensive, though their price depends on the material they are made from. Besides being made from plant fibres, the use-and-throw plates are also made from plastic and Styrofoam. Also, while plain white paper plates are the most popular, you can get them in a variety of colours, sizes, and designs.

In addition to their value-add to a picnic, paper plates have been put to various other creative purposes, thanks to their malleable nature and dexterity. In addition to serving a meal, paper plates are a staple in every kid's arts and crafts class. Here's how you can use them to nudge kids into budding makers.

PAPERCRAFT SPACECRAFT

Dennis is one of the most prolific authors at Instructables.com, and never misses an opportunity to participate in the site's regular themed contests: "Somehow I got hooked on creating new projects for every contest that came along, each one trying to surpass the last, a vicious cycle in over-achievement that calls on all your maker skills, and the tools and materials at hand." When the papercraft contest came along, he leaned on his geeky childhood, of watching shows like *Star Trek* and *Lost in Space*, to fashion a spacecraft from paper plates. He decided to create a model of the Jupiter 2 spacecraft from *Lost in Space*, perhaps because the paper plates are already in the shape of its saucer section. Dennis used

9" and 6" paper plates, and his Instructables lays down the exact process to transform them into the top and bottom sections of the spacecraft's saucer, complete with the skirt for the saucer's engine ring. After gluing the two together, along with a bottle cap for the dome, he cuts out the windows and equipment hatches, and uses folded pieces of paper to create the landing gear struts. ◻

> **" HE LEANED ON HIS GEEKY CHILDHOOD OF WATCHING SHOWS LIKE STAR TREK AND LOST IN SPACE TO FASHION A SPACECRAFT "**

Project Maker
DENNIS LOUIE

Project Link
hsmag.cc/jysmKb

Left ⬉
The spacecraft is lit with an Adafruit FLORA and a pair of NeoPixel rings that run the basic goggles sketch, which incidentally provides the same light animation as the original Jupiter 2

FEATURE

PAPER PLATE
MARBLE TRACK

To fuel the need for speed of her four boys, maker mom Sarah has been building marble tracks using everything from paper towel tubes to pool noodles. Fresh out of ideas, she was about to buy the boys a wooden set when she saw an article in a magazine that created tracks

Project Maker
SARAH DEES

Project Link
hsmag.cc/zvsGzg

—

> " SHE CREATED AN ELABORATE TRACK THAT WAS A LOT STEEPER – TO COMPENSATE, SHE ADDED GUARDRAILS ON THE TURNS "

using the rims from the paper plates. It took her a little time to adapt the idea into a track of her own. Sarah suggests using plates that have a design on them and a smooth rim around the edge. With a little trial and error, she created an elaborate track that was a lot steeper than the one in the magazine. To compensate, she added guardrails on the turns by turning the paper plate rims in the opposite direction. Sarah suggests building the track from the ground up and that the less steep it is, the easier it'll be to keep the marble under control. She also placed a lid from a jar at the bottom of the track to keep the marble from rolling onto the floor where her toddler could get to it. ◻

Below ◈
While the article suggests using different-sized paper towel rolls as supports, Sarah hot-glued her track onto cardboard supports

PAPER PLATE
RACER

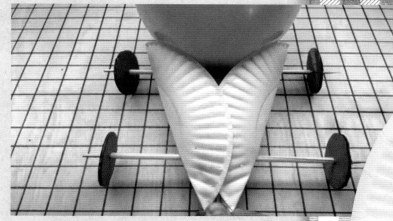

As the co-founder of DIY.org, Isaiah Saxon is well-known in the maker community. One of the many hacks the filmmaker has posted on the website is a paper plate rocket car that's propelled by a balloon. Isaiah begins by folding a paper plate into the body of the racer. He then uses some cardboard to cut out wheels and carefully places a couple of BBQ skewers into the centre to create a set of axles for the racer. The axles are taped to the bottom of the paper plate. To propel the contraption, he cuts the open end of a balloon and tapes a bunch of straws inside it to create a tailpipe. This tailpipe is then taped to the paper plate. To put the racer in motion, simply inflate the balloon through the tailpipe and let it rip in an open space. ◻

Project Maker
ISAIAH SAXON

Project Link
hsmag.cc/EMCvsL
—

Above ◈
Many characters in Isaiah's animation films indulge in some sort of DIY hacks. You can view his videos at isaiahsaxon.com

PAPER PLATE
SKELETON

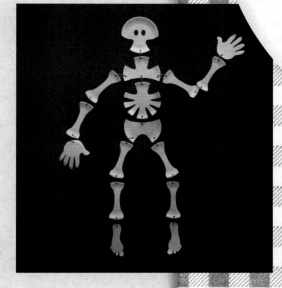

Although Halloween's come and gone, former Instructables employee Karen has an easy and fun festive-themed craft – a paper plate skeleton that's not anatomically accurate, but can be easily replicated by kids with minimal supervision and guidance. She used some paper plates that were 6″ in diameter and some string to piece together a skeleton. Her Instructable has images of all the different body parts that she's cut out from the paper plates, including a skull, a shoulder, a ribcage, a hip, a couple of hands and feet, and eight bones. You can trace her designs into paper plates – but for the hands and feet, she suggests tracing your own. After cutting out the

Project Maker
KAREN HOWARD

Project Link
hsmag.cc/KIZVWa

Right ⬈
The skeleton is easily customisable. Beef it up with more bones, or add a dash of colour

pieces, punch holes as shown in her images, and tie them together using a piece of string or twine. Karen suggests using some tape to hang the completed skeleton on a wall in any position you want. ◻

CANDLES

Ignite your creativity with this blazingly simple mantelpiece accessory

Mayank Sharma

🐦 @geekybodhi

Mayank is a Padawan maker with an irrational fear of drills. He likes to replicate electronic builds, and gets a kick out of hacking everyday objects creatively.

andles, in one form or another, have been used as a source of illumination for millions of years across the world. While the ancient Egyptians used wickless torches to continue to be productive after sunset, the true design ancestors of the modern-day candle can be traced back to Rome of 500 BCE. The earliest candles found in Europe used an unwound strand of twine or rolled papyrus, while the ancient Chinese used rolled rice paper for the wick. The choice of wax also varied, since it came from the flora and fauna prevalent in the respective regions. Candles in ancient Chinese excavations contained whale fat, and some in India were made by boiling cinnamon and yak butter. Beeswax though has been one of the most popular ingredients of the candle, which continues to this day.

Instead of being the primary source of illumination, candles have always been somewhat of a novelty item that have often been given as gifts and used in religious ceremonies across cultures and time. Candles have also been used for keeping track of time. King Alfred of Wessex famously used a candle clock that took four hours to burn, and had marks along the sides for every hour. These candle clocks helped track time when the sun wasn't visible; during the night or on a cloudy day.

Before manufacturing candles became an industry, it was a profession. Between the 5th and the 15th centuries, tradesmen known as chandlers went door-to-door across Europe to make candles from the kitchen fats that were saved for this purpose. These tallow candles were cheap but gave off a strong odour. On the other end of the scale were the beeswax candles that had a pleasant smell, but were expensive and usually reserved for the nobility. Over the next few hundred years, the only thing that changed about the candles was the source of the wax.

In 1834, Manchester-based inventor Joseph Morgan patented a device to produce moulded candles. His revolutionary candle-making machine used a cylinder with a movable piston to eject the candles as they solidified, and could produce about 1500 candles every hour. At the same time, a few chemists were working to distil paraffin, and by 1850 it became the go-to material for producing inexpensive, odourless candles. It wasn't long before distilling kerosene became commercially viable. Kerosene proved to be an excellent fuel for lamps and sounded the death knell for the paraffin candle as a source of light. Since then, candles have primarily been used as decorative items and a source of creative inspiration for intrepid makers.

SMARTPHONE CHARGER

Project Maker
DAVID MATTIASSON

Project Link
hsmag.cc/ckduyf

Left ⬈
David has another project in which he used a cheap Peltier element to drive a small fan

 Ithough hiking enthusiast David Mattiasson never forgets to pack spare batteries and solar chargers, they fail him when he needs them most, as it's usually overcast in Sweden. Since he always has a burner of some sort in his backpack, he decided to use a Peltier element to create a thermo-electric charger that produces electricity from heat. The temperature difference between the cold and the hot sides of the Peltier element generates an electric voltage that David uses to charge his phone. He has used the device to generate electricity by heating it from a gas burner and even tea candles. The physical construction of David's device is also pretty impressive. To transport away all heat, while keeping the contraption portable and light, David used a small heat-sink and leveraged some of the generated electricity to run a small fan to cool the device. He also used a couple of heat-insulated washers to block the heat from transferring to the other side. David has detailed the process for duplicating his device, and you'll also find some metrics on his Instructables page. ▫

FEATURE

CANDLE-POWERED CAROUSEL

Project Maker
TEISHA ROWLAND
Project Link
hsmag.cc/RaNQQr

Minecraft fan and a science educator, Teisha is always looking for ways to use her interests to spark scientific interest in youngsters. A science writer and author of the 'Biology Bytes' books, she is always looking to make projects that require things that can mostly just be found around the house: "I've also always been fascinated by the idea that heat can cause things to move (and fly!) – like paper balloon lanterns – but it can be challenging to demonstrate this physics principle, because the materials typically need to be very lightweight."

> **" I'VE ALSO ALWAYS BEEN FASCINATED BY THE IDEA THAT HEAT CAN CAUSE THINGS TO MOVE (AND FLY!) "**

Inspired by the design of the wooden Christmas pyramids, Teisha decided to create a Minecraft-themed paper carousel that's powered by candles. The heat from the candles causes blades at the top of the pyramid to rotate, which nicely demonstrates the involved thermodynamics. The most crucial elements of the project are the paper blades, and Teisha has shared the PDF of the design that you can print and cut out, as per her detailed explanations. She also handholds you through the process of constructing the other bits, including the platform and the figurines, and the final assembly as well. □

Left ◩
Since paper is highly flammable, Teisha advises you to maintain a distance of at least 12cm between the paper blades and the candles

SCULPTURED CANDLE

Looking for uniquely shaped designer candles at the price of the regular ones? Mae Berry shows you how to customise run-of-the-mill candles into very unique designer candles in the shape of a dragon, with some everyday knick-knacks. She shares a couple of ways to use hot water to soften a 12" tapered candle, which she then simply twists to create the body of the dragon. She splits one end of the candle with a knife to create the dragon's open mouth, and gouges out holes for the eyes. The limbs are created with a couple of 6" candles that are shaped in the same manner as the body. Mae explains the construction in her Instructables, along with lots of images to help you replicate every step of the process. The level of detail in her sculptured candle is absolutely amazing and will surely impress anybody you show it to. Just make sure you take adequate precautions when handling the hot water. □

Project Maker
MAE BERRY

Project Link
hsmag.cc/CSXqUO

Left ←
The candle has wicks at both ends – once the body is used, lop off the head and light the whiskers

Below ↓
The last hack has a bonus science fact: cold nitrogen gas is denser than air and puts out the candle before the liquid can get to it

PUT OUT A CANDLE WITH PHYSICS

You can put out a candle by blowing on it, but that won't teach you much. Dianna Cowern shows you various ways you can snuff out a candle flame and, in the process, learn a thing or two about the physics involved. Dianna is the host of the Physics Girl YouTube channel, which tries to creatively engage students in science, technology, engineering, and maths (STEM). In this video, she puts out the candles using different mechanisms, each of which demonstrates a different principle of physics, such as fluid dynamics, airflow, thermodynamics, and others. Our favourite is the one in which she creates carbon dioxide from baking-soda and vinegar, transfers the colourless gas into another cup, and dumps the seemingly empty cup over the flame to rob it of one of the key ingredients of fire, oxygen, and extinguish the flame. □

Project Maker
DIANNA COWERN

Project Link
hsmag.cc/DjbjEa

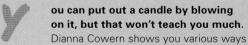

SYRINGES

Inject a dose of creativity into your builds

Mayank Sharma

🐦 @geekybodhi

Mayank is a Padawan maker with an irrational fear of drills. He likes to replicate electronic builds, and gets a kick out of hacking everyday objects creatively.

The projects in the following pages all use needleless syringes. In case you can't find these at the local pharmacy, make sure you carefully remove and dispose of the needle from the disposable syringes. Also, reusing syringes (even for non-medical uses) is a bad idea, so make sure you always use fresh syringes.

No one has fond childhood memories of syringes, but no matter how many kids shriek at its sight, the syringe is an indispensable piece of medical equipment. In terms of its build, the device is a simple reciprocating pump that has a plunger that fits cosily inside a cylindrical barrel. When pulled, the plunger creates suction – and exerts pressure when it is pushed. With these two actions, syringes are used to either inject or extract fluids through the skin, flesh, or veins.

While syringes these days are mostly used to administer or inject medicine, they were actually designed as extraction devices to remove harmful substances like pus and poison. The exact date of the invention of the syringe has been lost in the annals of time, but there are records of Hero of Alexandria, the Roman mathematician and engineer, describing the construction of a syringe-like device he called a 'pyoulcus', which roughly translates to 'pus puller'. Perhaps the oldest record of the use of syringes dates back to the 1st century CE. Aulus Cornelius Celsus, in his medical encyclopedia titled *De Medicina*, described procedures to clean wounds and remove pus using a syringe. The early syringes were made out of either metal or glass.

The 9th century ophthalmologist Ammar ibn Ali al-Mawsili devised a method to use a glass syringe to suck cataracts from the eyes. But these early syringes were expensive, were used repeatedly and, despite regular sterilisation, led to the transmission of diseases. Also, while they were good at extracting things, it was the invention of the hypodermic needle that allowed doctors to inject liquids into something other than pre-existing openings.

The first truly disposable syringes that were produced in large quantities were designed by James T Greeley in 1912. These were collapsible tubes made of tin (a bit like superglue or toothpaste tubes) that contained a certain amount of morphine for subcutaneous injection on the battlefield, and were very useful in World War I. However, it took several more decades before disposable hypodermic syringes entered the mainstream. Thanks to penicillin clogging up the glass syringes over repeated use, Charles Rothauser used polyethylene to make one that was cheap enough to be discarded after a single use.

Syringes have been a vital part of the medical field, but they do have non-medical applications as well – they have enough physics to be put to interesting uses.

SYRINGE-POWERED JCB

Project Maker
AKASH VAGHANI

Project Link
hsmag.cc/ihkEYR

Left ◧
The JCB sits on a crawler that's propelled by a couple of DC motors for forward and backward movements

We've had our fair share of RC cars, but none was as cool as this fully functional JCB. A major part of the build involves assembling the body of the digger, using pieces of cardboard and hot-gluing them together. The crux of the project, however, is its two hydraulic arms that move the bucket and lift the loader, and are made using syringes. Akash's system requires a total of four syringes. One is attached to the bulldozer's loader arm, and another to the bucket. The other two are used to control their movements. For the hydraulics, a pair of syringes are connected via a pipe and filled with coloured water. When you push the plunger on the control syringes, it'll push the plunger on the other that's connected to the bucket and the loader arm of the JCB. Inversely, when you pull the controller plunger, it'll pull the one connected to the JCB. You can now use them to mimic the motions of a real JCB digger. □

DIY VACUUM CHAMBER

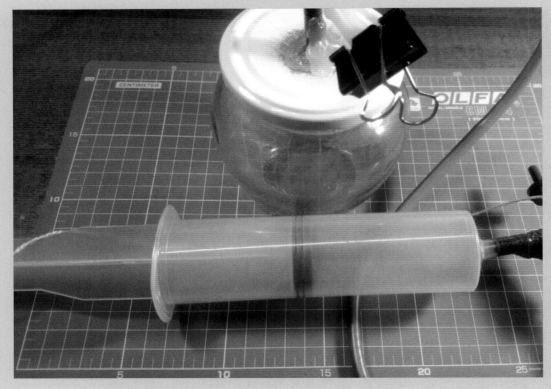

Project Maker
ALEXANDER MAYOROV

Project Link
hsmag.cc/RIufXR

Left ◈
You can use the pump to create enough pressure to inflate balloons and boil some water

T **his one is for all you STEM enthusiasts.** If you need a portable DIY vacuum chamber to demonstrate what happens in a low-pressure environment, Alexander shows you how to build a hand vacuum pump that can create a low pressure inside small chambers. He uses a 150 ml syringe, a couple of two-way valves and T-connectors, a drop counter, and a length of silicone tube. If you can't find the tiny valves and T-connectors in your local hardware store, Alexander suggests looking for them in a pet shop that stocks or repairs aquariums. He explains the entire process of using these bits and pieces to assemble the pump, which is fixed atop a glass jar with a lid. Since it's unsafe to open the vacuum chamber while it's vacuumised, Alexander shows you how to create a vacuum release valve using a paper clip. You can now use the pump to create a low pressure of up to 0.2 bar, which is enough to demonstrate the characteristics of a low-pressure environment. ▫

DIY FOOD SLICER

If you love cooking, Denny has a useful hack for you. He shows you how to use a syringe and a steel wire to julienne vegetables in a jiffy. Take a syringe with a detachable plunger and a wide barrel, lop off its top, and then sand the rough edges. Now, wrap electric tape around the freshly sanded top and use a calliper to mark equidistant holes around the syringe. Now, use a drill to make the holes and carefully weave a steel wire through them, forming a checked pattern. This is the most time-consuming and error-prone step of the entire process. Once you've woven through all the holes, loop the wire across the barrel, and make sure it's firmly in place before snipping off the rest. That's all there is to it. Now, remove the plunger, put in a piece of vegetable or sausage, and press the plunger to force it through the woven steel wire. □

Project Maker
DENNY

Project Link
hsmag.cc/CnmpWI

Above ◈
At the beginning of the video, Denny shows you how to use a syringe to create a height-adjustable microphone stand

UNIVERSAL GRIPPER

Amazed by videos of the universal gripper made by iRobot, Charles Ford decided to build his own: "I kept thinking how useful it could be for people who can not easily deal with pills. The idea that a robot arm could pick up small items and deliver them to someone bedridden or physically handicapped motivated me." To complete the build with less power and complexity, he thought of using syringes, inspired by their use by special make-up effects creator Rick Baker for *An American Werewolf in London*. Charles has a detailed list of parts in his Instructable, and explains all the construction steps. He went through several revisions and has illustrated his progress with photos and videos. His universal gripper can pick up very irregular or smooth objects easily, despite the fact that it uses off-the-shelf syringes instead of expensive vacuum pumps and reservoirs. □

Project Maker
CHARLES FORD

Project Link
hsmag.cc/cEssbC

Left ◈
Charles got the large syringe from a veterinary supplier who was taken aback when told that it wasn't for his horse, but his robot!

How I Made

AN ACCESSIBLE GAMING STEERING WHEEL

Making games that more people can play

By **Andrew Lewis**

My first experience with a racing wheel was in the late nineties, when force feedback technology had just been released, and Need for Speed 3 was my favourite game. The wheel made the game so much better, but my wheel connected to the computer via a serial connection, rather than the new USB-style connection that was becoming popular. When I upgraded my operating system to the latest version, the drivers for the wheel were no longer compatible. With the technology now useless to me, my curiosity got the better of my judgement, and I decided to pull apart the controller, and see how everything worked. To my surprise, the controller was nowhere near as complicated as I had expected it to be. Even more surprisingly, I spotted an unsoldered jumper pad on the

Above
A modified race wheel pedal controller, fitted to a race wheel

control board inside the wheel, with four connections labelled USB. An hour, and an inch of solder later, I was playing my favourite game again, with a USB-compatible race wheel.

Twenty years later, I had a conversation with my friend and colleague Chris Power. Chris is the vice-president of The AbleGamers Charity, and it turned out that my early experience with race wheels might be useful for adapting them for people who aren't able to use them in their default configuration. Like most game controllers, a race wheel (and an actual car) is designed for someone who has full control over all of their limbs. While cars can often be adapted to suit individual needs, with hand controls or levers instead of pedals, the same can't always be said of game controllers. Chris suggested that I might be able to help on one of their projects, and I was very happy to donate some time to a worthy cause.

How the wheel works

This steering wheel modification replaces the accelerator and brake pedals of a Logitech race wheel, with hand controls positioned on the wheel itself. The main accelerator and brake are interlocked, so that pushing the brake lever down will also reduce the acceleration. This interlocking action can be disabled by removing the grub screw from the acceleration gear. In addition to a brake and accelerator lever, there are three additional buttons. These buttons are programmed to mimic certain common driving pedal states when they are pressed down:

Turbo button
Sets maximum acceleration and zero brakes. In the default configuration, this button is located at the top of the column of buttons on the panel.

Coast button
Sets accelerator and brake to zero. In the default configuration, this button is located in the middle of the column of buttons on the panel.

Emergency Stop
Sets the accelerator to zero and brake to maximum. In the default configuration, this button is located at the bottom of the column of buttons on the panel.

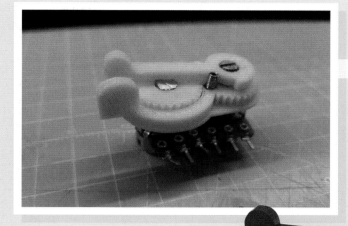

Left
Final version of the lever design, before mounting in the control panel

Below
3D model of the modified lever design, after discussion with user

their perspective, and seeing whether they have any ideas how that problem could be solved.

The problem was that the pedals weren't accessible to the user, and it wasn't possible to control the throttle and brakes effectively using the existing buttons on the wheel. I came up with the idea of a linked lever hand control, and I presented this to the user as a possible solution. After some more discussion, we decided that the lever control alone wouldn't be enough to solve the problem, so I added some more buttons to the design and asked whether the user liked my idea. I had some more conversations with Chris and the other volunteers at The AbleGamers Charity. We picked up on some issues that would make the project more flexible, and we came up with a workable specification that covered the installation, calibration, and use of the race wheel to play games. We wanted the wheel to be easy to install, easy to calibrate on different machines, and comfortable to use when gaming.

INVESTIGATING THE HARDWARE
Now that I had a design, I could look at the hardware and figure out how to implement

DEFINING THE PROBLEM
The temptation with any project is to dive in and prototype things. From experience, I know this is a bad idea. The first step in solving a problem should always be to define the problem clearly. In this situation, defining the problem meant having a conversation with the end-user, finding out the problem they're experiencing from

it. The pedals on the race wheel were connected to the main wheel via a 9-pin D-Type connector, and I started to probe the pinouts from the pedals with my multimeter. I figured out that (as I had suspected) the pedal unit was entirely passive, and the pedals were just plastic shapes connected to potentiometers. The output from the potentiometers ran directly into the 9-pin connector on the race wheel. I recorded the maximum and minimum resistance of each pedal, and soon realised that the amount of movement needed would be too much to just move the potentiometers directly onto the wheel and connect levers to them. I could try to modify the movement range using gears, or I could use electronics to map the values from a different set of potentiometers to match the output range of the pedals. An electronic solution meant that I would be able to integrate button controls more easily so, after determining that the race wheel supplied 3.3 V to the pedals, I ordered an Arduino Pro Mini 3.3 V and waited for delivery.

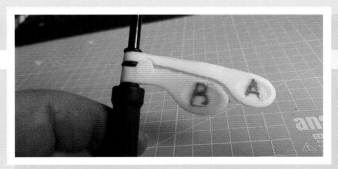

Left
The original lever concept described to the end-user

Using multiple MCP4725 boards

When using MCP4725 boards, the I²C address can usually be changed between two values by moving a solder jumper on the face of the board. This must be done on one of the MCP4725 boards so that each board has a different address when connected to the Arduino. You must also disconnect the pullup resistors from one of the MCP4725 boards by cutting the tracks to the resistors.

Above
The back of the panel during assembly, showing the switch and lever connections, and the MCP4725 boards ready to be connected to the Arduino Pro Mini

BUILDING THE SOLUTION

With some electronics hardware on the way, I started designing the lever assembly in Rhino 3D, and 3D-printing the parts. Now that I knew I was using electronics to convert the resistance levels, I had the freedom to choose good quality, compact potentiometers and start deciding what other components I would need to make the modification work as expected. The first problem to overcome was that the Arduino needed to mimic the effect of the pedals in a way that the race wheel could understand. I could have solved this mechanically by using servos connected to potentiometers, but I decided that more moving parts would increase the potential points of wear and failure. Instead, I set about designing a circuit that would simulate the effect of several potentiometers, using digital to analogue converters (DACs). A DAC takes a digital signal from the Arduino, and converts it into a voltage between 0V and 3.3V.

The MCP4725 is a common choice for this sort of application, so I connected two of them (one each for the brake and accelerator) to the Arduino, using the I²C system. The I²C address used on different brands MCP4725 boards can vary, so I made sure that I set a custom variable near the top of my Arduino code to make changing the address easy.

In addition to the throttle controls, the pedal modification also includes three buttons that mimic common foot pedal actions. The Turbo button sets the accelerator to maximum and the brake to zero. The Coast button sets both the accelerator and brake to zero. The Emergency Stop button turns the accelerator to zero, and the brake to maximum.

The next problem I dealt with was finding an easy way to calibrate the system. The modification to the race wheel

I set about designing a circuit that would simulate the effect of several potentiometers

would work if I hard-coded the conversion values for the pedals and levers, but it's possible that different race wheels will use different configurations. I wanted my modification to be easy to adapt and use on multiple devices, so I developed a software calibration system that only needs to be set up when the system is first fitted. The person calibrating the system attaches the pedals to the control box and flicks the calibration switch. Then, the user sets the pedals and levers to their lowest positions and pushes the Emergency Stop button. The pedal and lever positions are reversed, and the user pushes the Turbo button and turns off the calibration switch. This gives the Arduino everything it needs to detect the maximum and minimum values of the levers and pedals, and set the DAC values correctly when the user moves a lever or pushes a button. The values are stored in the Arduino's EEPROM, so it isn't necessary to recalibrate if the race wheel is unplugged from the computer.

FINAL TOUCHES

With the hardware working correctly, I started designing more presentable hardware. Up to this point, I'd been

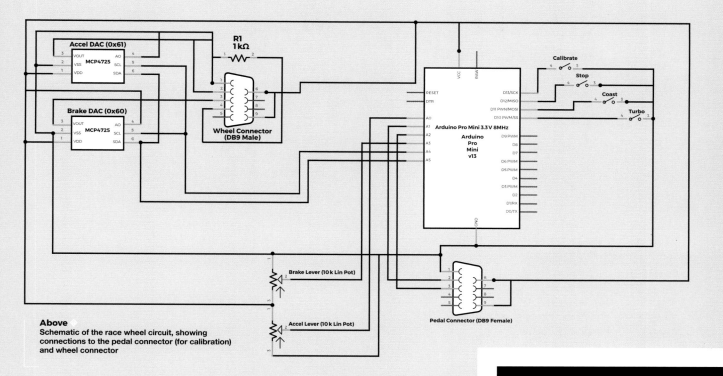

Above
Schematic of the race wheel circuit, showing connections to the pedal connector (for calibration) and wheel connector

using a piece of scrap plastic to mount the switches and levers onto the wheel, so I 3D-printed a suitable panel and a box to hold the electronics. I used 9-pin D-type connectors to join the wires, so that replacing them would be easy in the event of a problem. I also made proper documentation for the project. It's unlikely that I'd be the person who had to remake or install anything, so I made sure that all of my code, parts lists, and diagrams were available for the volunteers at AbleGamers to remake and modify

the project. I didn't need to make a final assembly of the project myself, because I knew each of the individual parts worked fine. It was more useful to hand the project over to AbleGamers, who could build a complete version from the instructions that I provided them with, and help find any mistakes in my notes. ▫

Below
The 3D-printed enclosures designed for the final assembly. I produced enclosures for the control panel, Arduino (with calibration switch), and also for the 9-pin connectors

The AbleGamers Charity

Founded in 2004, AbleGamers is a 501(C)(3) nonprofit charity that enables people with disabilities to play the games they love, or to play games for the very first time. The AbleGamers headquarters in West Virginia are home to Songbird Studio, a workshop with equipment for engineering and making assistive devices. AbleGamers has helped thousands of gamers through its grant program, by allowing people with disabilities to apply to receive free, custom-made assistive technology. The AbleGamers Expansion Pack program goes even further, by supplying hospitals and long-term living facilities with entire gaming room setups that include the assistive technology patients will need to game.

From working with the University of York on cutting-edge research, inventing innovative technologies with the world's best engineers, and deploying free industry-standard resources like **accessible.games**, AbleGamers continues to impact real change to a multibillion-dollar industry to be a more inclusive virtual playground.

Contact: ablegamers.org

FEATURE

How I Made
A VERTICAL PLOTTER

Building my own drawing machine

By **John Proudlock**

A few years ago, I ended up with a spare Raspberry Pi, and I set about looking for a project to put it to good use. I had a bit of a root around on the internet for inspiration. There was no shortage of ideas, but I wanted something that involved movement – I'd never used the GPIO ports on a Pi. I also wanted something that needed a bit of simple coding – I'd not coded for years, and I wanted to make something that would have some vaguely artistic output.

And that's why plotters caught my eye. I've never been very good at drawing. I always wondered if I could build a machine to help me… It turns out I could. The basic principles are fairly straightforward, and this article explains them.

You'll probably have seen commercial plotters. Typically, the paper is set horizontally, and an armature moves a pen over the surface to draw. There's a much simpler version, known as a vertical plotter

(depending on the creator, they might also be called v-plotter, polar plotter, or polargraph). Search for images online, and you'll quickly see how these devices work – but I've marked up the photo of the one I built opposite, and its main parts:

A. Paper attached vertically to a wall (on the plotter easel)

B. At the top-left and right corners, there are two motors; the drive shafts point out into the room

C. Pen holder, suspended on a toothed belt running between the two motors

D. When the motors move, the pen is dragged across the surface of the paper, creating an image

E. Toothed belt moving over the motor sprockets

F. Weights (steel chain) to counter the weight of the pen holder

G. Control electronics with small output display

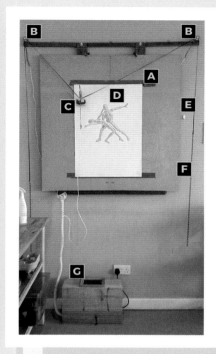

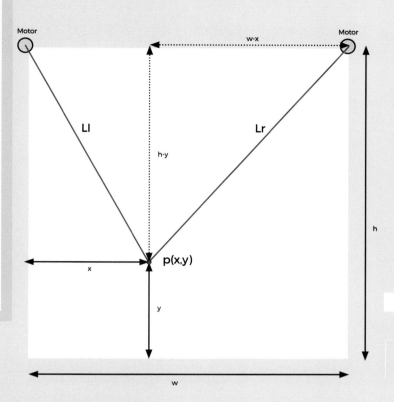

Left
Understanding
the maths

The control electronics:

A. The power supply unit (from an old desktop PC)

B. PSU breakout board

C. Raspberry Pi B+ and GPIO breakout

D. 7-inch HDMI screen (the back of)

E. Breadboard
- Stepper motor drivers to left and right motors
- Drive for the pen lift servo

F. Cooling fan for motor driver circuits (normally mounted over the breadboard)

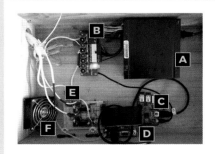

Left
The control
electronics

V-PLOTTER MATHEMATICS

I start new projects by working on what I think will be the hardest part of the project for me to complete. That way, if the project is beyond me, I find that out early on and abandon it! For the plotter project, I wanted to make sure I could understand the underlying mathematics. I sketched out a simple model of a plotter.

I wanted to work in Cartesian coordinates, so if I want the pen at point p(x,y) how long would the belts from the left motor (Ll) and the right motor (Lr) need to be? To know that, I also need to know the height of the paper (h) and the width between the two motor spindles (w).

The v-plotter makes two right-angled triangles, and the values can all be calculated using Pythagoras' theorem:

$$Ll = \sqrt{(x^2 + (h-y)^2)}$$
$$Lr = \sqrt{((w-x)^2 + (h-y)^2)}$$

So, if the plotter is 500 mm square, h=500, w=500. And, if I want the pen to be at p(150,200), then x=150 and y=200 so the belt lengths can be calculated as:

$$Ll = \sqrt{(150^2 + 300^2)} = 335.4$$
$$Lr = \sqrt{(350^2 + 300^2)} = 461.0$$

I then started to think about moving the pen across the paper to draw a line. I bought a pair of NEMA stepper motors which are driven by electrical pulses (that would come from the Pi's GPIO). For each pulse, the motor would turn 1.8 degrees, so 360/1.8 = 200 steps per full rotation of the motor shaft.

I then dipped into the world of 3D printers to get a GT2 toothed pulley for the motor shaft, and decided I'd use a GT2 toothed belt to run over each pulley and suspend the pen. This had the advantage that the pulley wouldn't slip or stretch (unlike string). The sprockets I found

Figure 1
I never did work out
the maths to correct
distortion in long line

Left
Using a GT5 sprocket
to precisely control
the plotter

had a circumference of 40 mm. This means one rotation of the motor would take 200 pulses, and would move the pen 40 mm closer/further from the motor. You can see the left-hand motor setup in the image above.

In the example before, if the pen is at point A(150,200) we know Ll = 335.4 mm and Lr = 461.0 mm.

If I want to move the pen to Point B (350,450) then we need to make Ll = 353.6 mm and Lr = 158.1 mm.

Just considering the right-hand motor and Lr, the change in Lr is 461 − 158.1 = 302.9 mm. This requires 302.9/40 = 7.57 rotations of the motor spindle.

As a single rotation requires 200 pulses, we can calculate that the right-hand motor must receive 200 × 7.57 = 1514 pulses.

You can use the same working to calculate the pulses needed to set the new length of the left-hand belt, Ll (the answer is 91 pulses).

HARDWARE – SETTING UP THE MOTORS

Confident that the underlying mathematics was all manageable, I moved onto the hardware and getting the two motors powered up. The Pi GPIO would provide the signal (the pulses mentioned earlier) as a square wave, but the Pi can't provide the current needed to drive the motors, so you need some additional circuitry to boost the amps.

Firstly, I connected the GPIO from the Pi to a breadboard so that the signals from individual pins were accessible. It was also easier to get pre-built stepper motor driver circuits to save time, and I found the (excellent) 'EasyDriver – Stepper Motor

Driver' circuits. I soldered on the headers so the two driver circuits would plug into the breadboard.

I connected the four power lines to each stepper motor and powered up the driver circuits. The motor spindle went from being loose to fixed firm (stepper motors draw current and lock without an input signal).

I always intended to transfer the breadboard circuit onto stripboard, but never got around to it. By using short wires, close to the board, the breadboard solution has been robust enough to work reliably.

There were two inputs required to make the motor move: the 'drive' signal, which would be a square wave where each pulse would move the motor spindle 1.8 degrees, and a 'direction' signal as the motors would need to rotate back and forth depending on the direction of pen line required. Armed with this knowledge, it was time to write the first code for the project.

> By using short wires, close to the board, the breadboard solution has been robust enough to work reliably

Above
Portraits, starting
with the basics

SOFTWARE – DRAWING A LINE

I used Python and worked in the IDLE environment to simply set a GPIO pin high, then low, 50 times a second. I did this for four seconds, and I got a motor spindle to turn one rotation clockwise. I then set the 'direction' pin to high, and repeated the four-second cycle. The spindle turned one rotation anti-clockwise.

From there, I duplicated the code for the other motor and clamped each motor to a shelf: one on the left, one on the right. Under the shelf, I propped up a whiteboard. I took the toothed GT2 belt and draped it across the two motors. I gaffer-taped an old takeaway carton in the middle of the belt and glued a dry marker to it. With the left motor turning anti-clockwise, and the right motor turning clockwise, the pen was dragged up the board and made a neat vertical line.

I wasn't going to meet any artistic needs with a single vertical line, though, so I wrote a function that would be passed an (x,y) location and would move the pen from its current location to the one I requested.

The main steps to this are:

- Knowing your current pen position

- Knowing the position you want to move to

- Calculating the change in the left belt, and the right belt

- Calculating how many steps are required for each motor to make the change in belt lengths

- Moving the motors

I had the maths for this from earlier in the project, so coded it up. The first stumble was that if you move one motor, then the other, you don't get a straight line to the destination, you get a sort of L shape. So you have to use two processor threads, one for each motor. Both motors have to start moving at the same instant.

I found that I needed a more sophisticated development environment than IDLE and moved the project into PyCharm. The PyCharm debugging and refactoring tools were essential as the code got more sophisticated.

Unfortunately, that still left me with a kinked line because (unless it's a vertical line in the centre of the paper) the L belt change and the R belt change are different lengths, so one motor finishes first and you end up, again, with a curved L shape instead of a straight line.

To fix this, the motors have to run for the same period of time, starting together and

Version control

As the code became more complex, I learnt to use the very basics of Git, a code repository tool. This made it easy to go back to known 'safe' versions of the software if I needed to. I went on to use Bitbucket (an online repository). This let me work on the software on a desktop PC, then 'push' the code up to Bitbucket. On the Pi (controlling the plotter), I'd simply 'pull' down the latest version of code that I wanted to use.

Right
Portraits, looking basic

finishing together. To do this, you need to have set your fastest motor speed. I found 50 pulses per second worked, and used that as the 'maximum motor speed':

- Identify which belt length change is the biggest. In the example from earlier, this is the change to Lr, which requires 1514.32 pulses. At 50 pulses per second, this will take 30.29 seconds

- The left motor must change the Ll length by just 91 pulses. These must be delivered through the 32.29 seconds that the right-hand motor is running, which requires a pulse rate of 91/32.29 = 3 pulses per second

I coded this so that both motors ran for equal periods and… it worked, mostly. Short lines of about 5 mm appeared straight. Long horizontal lines would curve slightly, like they were sagging. You can see this in **Figure 1** from early in the plotter's development. The plotter is drawing squares of increasing size. As the horizontal lines get longer, you can see them dip in the middle. There was distortion on the vertical lines too, but it was not as noticeable.

The cause is that in my implementation, the pulse rate sent to the motors is constant, and each motor rotates at a fixed rpm as the pen moves. Instead, the motor rpm should vary constantly during pen motion. I haven't attempted to understand the mathematics required to fix this (I'm not sure I could), so instead, I found a workaround. Any long line is broken into segments, each no longer than 5 mm. In this way, a long line is drawn in several short segments. Although each segment

The plotter is drawing squares of increasing size

bows slightly, it's not visible and the line appears straight.

In this prototype version of the plotter, you can see that I'm plotting on a whiteboard and that the marker pen is held in an old takeaway container. The container holds my trusty Buddha paperweight to stabilise the pen.

I really wanted to draw something a little more inspiring than squares and started to experiment with drawing bitmap images, all with a single line. I started by considering a photograph I had and simplifying it into a small bitmap of 23×23 pixels.

I then mapped these pixel values to a 'darkness level' by equating black to a darkness level of 100 and white to level 0. In the plotter control code, I stored these values in a two-dimensional array. The plotter was then configured to represent each pixel in a 20 mm square and started with the pen in the top-left corner of the image. The latter is a white pixel (level 0), so the pen moves to a random point in the next adjacent pixel. If the next pixel is black (represented by 100), the pen must keep moving to random points in the pixel's square until 100 mm of line has been drawn. After that, it moves to the next adjacent square.

This simple algorithm would 'colour in' a 20 mm square with an amount of line represented by a pixel's darkness level. In this way, graduating tones would be

represented too. A mid-grey (value 50) would cause 50 mm of line to be drawn in a pixel's square.

Once coded up, I set the plotter running and waited to see how close the plot was to the image I had passed in.

With some squinting, I could just about make out the original image but could see that to make an engaging image, I would need a much higher resolution. This would need a bigger drawing space and a smaller pen tip. In any case, the shambolic 'tape a couple of motors onto a shelf' approach was not proving particularly reliable. This triggered the building of a simple easel that I could hang on the wall. In turn, I also upgraded the drawing algorithm to work in 5 mm squares per pixel, and this allowed the re-creation of 100-pixel images, and much better image creation.

WHAT NEXT...?

This is a 'weekend project' that's now been running for about four years, and some extensions I've found fun to add to the project are:

- Plotting bitmap images with a single line

- 'Generative Art', which basically means plotting the results of equations. It may sound dull, but the results can be stunning:
  - Plotting spirographs (the equations are scary-looking, but not that hard to replicate in code – even if you don't understand them)
  - Plotting fractals
  - Plotting modulo arithmetic

- Adding a mechanism to lift the pen off the paper

- Plotting G-code, an industry standard file format for CNC machines

- Drawing bitmap files with multiple colours

- Writing an emulator so you can predict how the plot will look on screen before you waste another set of pens and a sheet of paper

Each of these hacky endeavours would be an article in itself, but you can see the results from this plotter project at **instagram.com/inkylinesplots** and **inkylines.blogspot.com**.

How I Made

AN AUDIO EQUALISER TO HELP ME HEAR

Fighting back against the ravages of time

By Dermot Dobson

Many people (especially those of us in our fifth or sixth decade) struggle to hear voices clearly from TV programmes. You may have read in the press about the many complaints from viewers who were unable to pick out dialogue from some high-profile dramas recently. Voices are lost in bass-heavy music. This is caused by our ears losing frequency response as we age. Technically known as presbycusis (from the Greek words for 'old' and 'hearing'), the results can be dramatic. The problem isn't just that hearing goes, but that it goes differently depending on the frequency of the sound.

My hearing (as measured by an audiometrist when I got my hearing aids), is not too far off **Figure 1**, though there is a slight difference between left and right ears.

Above
Prototype layout of a single board. Frequencies increase from L to R

I have recently started using NHS digital hearing aids; they work well in environments with no echoes (I can now hear bird-song clearly while out walking – for the first time in many years), though they don't work well for TV use. While these hearing aids correct the age-related loss of middle and high frequencies needed to clearly hear voices 'naturally', most of the affordable types (including my NHS ones) have a 'behind-the-ear' microphone, which accentuates echoes from around the room, particularly when there is a wall close behind, with a TV directly in front. For a few years now, in deference to my wife's excellent hearing, I've been using a pair of wireless headphones fed from the TV – these have the disadvantage of not correcting the frequency response, though boosting overall sound levels. Of course, they also boost sound that doesn't

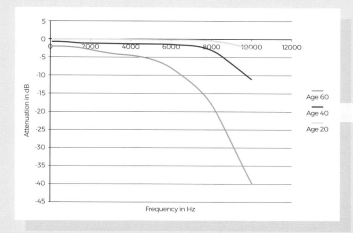

Figure 1
Common frequency-related hearing loss with age. X axis is frequency in Hz, Y axis is attenuation in dB for three different ages

Age 60
Age 40
Age 20

contribute to understanding dialogue clearly! In addition, by keeping sound levels high, I'm possibly causing further degradation in my hearing, due to the overall high volume sometimes required. While there is an option to use the 'T' setting on the hearing aids with an inductive loop hard-wired to the TV, I would then only hear both channels mixed down to mono.

Clearly, what is needed is a pair of wireless headphones with a way of adjusting the frequency response to more or less emulate hearing aids – but after a good look around, I find these are extremely expensive (and, looking somewhat flimsy, prone to expensive breakage too) or have no more than a bass boost – about the last thing that's useful to us sufferers.

> ## I still had trouble with higher-pitched voices

GET MAKING
For my first attempt at an equaliser, I hacked up a simple analogue three-band (bass, middle and treble) Baxandall-type tone control using a pair of op-amps, and inserted it into the 3.5 mm cable from TV to headphone transmitter. The result was significantly better, but I still had trouble with higher-pitched voices and louder music.

I clearly needed more specific control than this type of circuit could offer. Forty years ago I built a ten-band stereo graphic equaliser, that used inductors and capacitors for filters, but it was the size of an average domestic amplifier. Some of the inductors were quite large, and changing frequency bands wasn't easy. I needed something like this, but more compact and easier to tinker with.

What would I do differently in the future?
A few friends have expressed interest in getting one of these, having seen the prototype. For those, I will use full-sized stereo potentiometers, so that frequency response can be adjusted without diving into the box to adjust tiny presets. Initially, I opted for individual controls for left and right, as I could see that my ears are of slightly different frequency response. In practice, this proved to be insignificant, at least in my case.

I would also fit a changeover two-pole switch so the equaliser can be switched into and out of circuit instantly. I've found it very instructive for people with normal hearing to listen with the controls set to the *inverse* of the settings I use – they get an understanding of how those with hearing loss hear the world.

If making a number of them, a stereo PCB layout with a balance control would be the way to go. I'd also make provision for an optional buffer amplifier so it could drive wired headphones directly. I'd like to make this easier to construct so as many people as possible can benefit.

Enter the LA3600 series of chips; a five-band 'graphic equaliser' linear chip. Essentially, it has input and output buffers and five stages of 'gyrators' (a rather neat idea that uses op-amps and capacitors to emulate the bulky inductors that would usually be needed for the series of band-pass/band-stop audio filters that I used long ago). →

Right
The next iteration of the author is unaffected by time

Far Right
The next iteration of the equaliser with full-sized controls

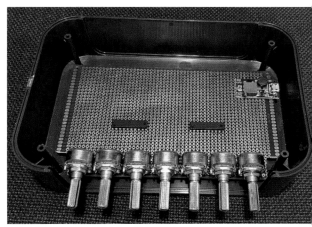

FEATURE

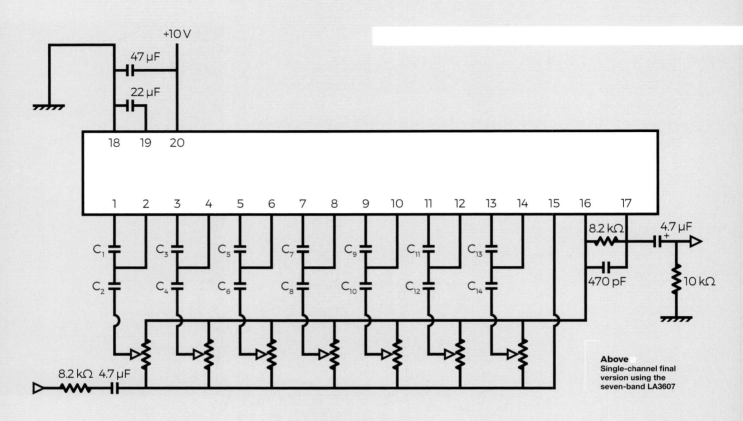

+10 V

47 µF

22 µF

18 19 20

1 2 3 4 5 6 7 8 9 10 11 12 13 14 15 16 17

8.2 kΩ

4.7 µF

C_1 C_3 C_5 C_7 C_9 C_{11} C_{13}

C_2 C_4 C_6 C_8 C_{10} C_{12} C_{14}

470 pF

10 kΩ

8.2 kΩ 4.7 µF

Above
Single-channel final
version using the
seven-band LA3607

The default values from the LA3600 datasheet provide filters at 108 Hz, 343 Hz, 1.08 kHz, 3.43 kHz, and 10.80 kHz – these values are appropriate for the intended domestic or car audio system use. This improved matters still further, but there wasn't enough selectivity to fully boost the voices, without increasing the other sounds. Fortunately, you can change this by switching in different capacitor values. I got better results by recalculating values to work only in the 800 Hz to 6 kHz range,

but there was no control over music in the lower registers.

I then started over with the LA3607 chip; similar to the 3600, but supports seven bands instead of five. This made a substantial difference, and I can now hear dialogue clearly, even on programmes where the sound is rather 'muddy'.

THE FINAL CIRCUIT

The default values from the LA3607 datasheet provide filters at 60 Hz, 150 Hz, 400 Hz, 1 kHz, 2.5 kHz, 6 kHz, and 15 kHz. I found the top frequency of little value in emphasising voices, so I recalculated the values to bring the top frequency down to a

Left
Go beyond 5 V
from USB with a
boost converter

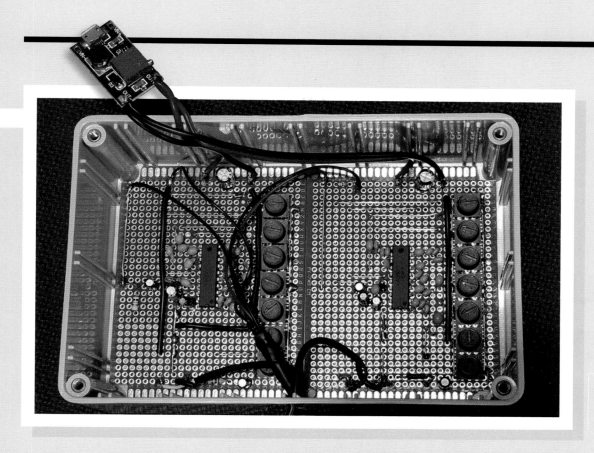

Left
Both boards
mounted, ready to
mount the USB boost
board on the lid

more useful point. I've found the best results (for my hearing) by selecting the following frequencies (which are approximate, bearing in mind the limited range of standard value capacitors that are easily available): 160 Hz, 345 Hz, 500 Hz, 915 Hz, 1.6 kHz, 3.4 kHz, and 6.25 kHz. These are a good balance between boosting voices while still being able to hear a reasonably full range of music.

You can experiment here: just keep the even-numbered capacitors of each filter to be 18 times (approx.) the value of the associated odd ones, and you can change the channel values/ spacing. You can see that the frequency scales linearly with the capacitor values.

Filter component values I used are:

160Hz –	C1:27nf,	C2:470nF
345Hz –	C3:12nF,	C4:220nF
500Hz –	C5:8.2nF,	C6:150nF
915Hz –	C7:4.7nF,	C8:82nF
1.6KHz–	C9:2.7nF,	C10:47nF
3.4KHz –	C11:1.2nF,	C12:22nF
6.25KHz –	C13:680pF,	C14:12nf

Depending on your requirements, you may care to experiment with some other values, perhaps skipping one or more of 160 Hz,

345 Hz, or 500 Hz and adding compensation for higher or lower frequencies:

75Hz –	56nF,	1000nF
5KHz –	820pF,	15nF
7.5KHz –	560pF,	10nF
11KHz –	390pF,	6.8nF
16KHz –	270pF,	4.7nF

Note that you will see a degree of interaction with filters when set to be closely spaced.

I also use an equaliser between my PC and speakers

For ease of use, I chose a standard USB charger for power, with a boost converter to get the correct voltage. If possible, I prefer using otherwise discarded stuff, and almost everyone has one or more old wall warts kicking around. The draw at 5V is only around 50 mA, measured by a USB power checker, so any type should do the trick.

My build came in at around £14 of parts (plus from my stock of components). The only potential pitfall in the build is making

sure you set the voltage correctly (I settled on 10V) *before* connecting to the load, because the boost converter modules come already set to some random value up to 24 V! I also use an equaliser between my PC and speakers; now I hear podcasts and other audio content to be far better.

Of course, the audio does sound rather strange to anyone with normal hearing; a bypass switch could make this a bit more useful for other people.

READY TO LISTEN

For flexibility, I chose to leave flying leads with in-line 3.5 mm plugs and sockets, as all of the cheap wireless headphones I have use those connectors. Follow the colours to ensure that you don't swap over left and right channels along the way.

The easiest way to set up the equaliser is by putting one earphone on one ear only, then adjusting the filters to suit. Most users will probably need a little 'cut' at the lower frequencies and a progressively increasing 'boost' at higher frequencies. Once complete, copying settings to the other channel is a good start, before doing any final small adjustments.

Arduinoflake

By **Jiří Praus** 🠒 hsmag.cc/wwRjLy

"**am a senior engineer for Samepage.io, and hardware enthusiast.** I started with a simple Arduino kit two years ago, and I fell in love with the platform. Now, I am having fun making free-form sculptures and electronics. My Arduinoflake has 30 LEDs interconnected by 0.8mm brass wire by the, so-called, dead bug method [where wires are soldered directly onto the upside-down integrated circuit] into the shape of a snowflake. It runs on Arduino Nano, and you can interact with it by the capacitive touch sensor. I wanted to build it as a toy for my daughter, but it turned out to be more – it's circuit art." □

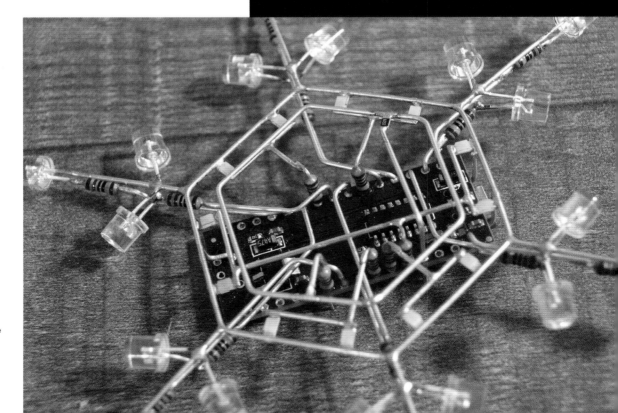

Right 🠒
This sculpture, like Kelly Heaton's bird, was entered into the Hackaday Circuit Sculpture Contest (hsmag.cc/whrlzk)

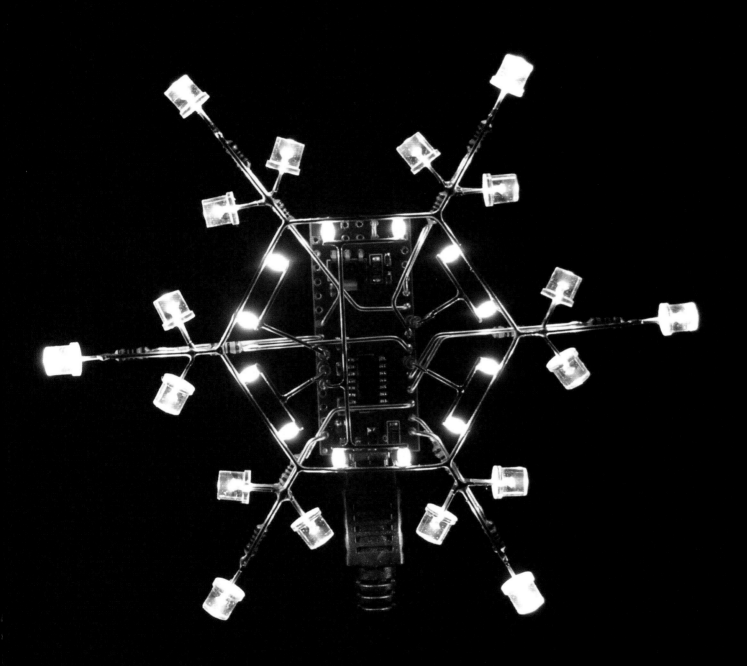

FEATURE ━━━━━━━━━━

As We Are

Ever wanted to see your face up in lights?

By **Andrew Gregory** 🐦 @AndrewGregory83

This is As We Are, a 14-foot tall interactive sculpture imagined by Matthew Mohr and built by DCL, a Boston-based company, specialising in fabricating really cool things.

It's built out of layers of contoured aluminium, covered in custom SANSI LED modules, comprising 850,000 individual LEDs. In the back of the head there's a photo booth, with 32 Raspberry Pis and Camera Modules, controlled by custom facial recognition software that finds your face, flattens it, and maps it onto the LEDs on the exterior of the sculpture, making your face twice as high as André the Giant.

As We Are is located in the Greater Columbus Convention Center in Columbus, Ohio; it was inspired when Matthew wanted a project to show off the diversity of the city.

This struck us as a brilliant project for a couple of reasons. The first is that it's a giant version of the Frank Sidebottom head. The second is that, at its heart, it's just some Raspberry Pis, Camera Modules and LED screens, meaning that it should be possible to make one of these yourself (OK, a much smaller, lower resolution one, but technically similar). Go forth and give it a go. ◻

↗ **hsmag.cc/eYjPdW**

Right ◈
Every subject in
As We Are is also a
participant – this is
really public art

Right ↗
During the day, the
sculpture faces into
the atrium. At night, it
rotates to face outward
to the street

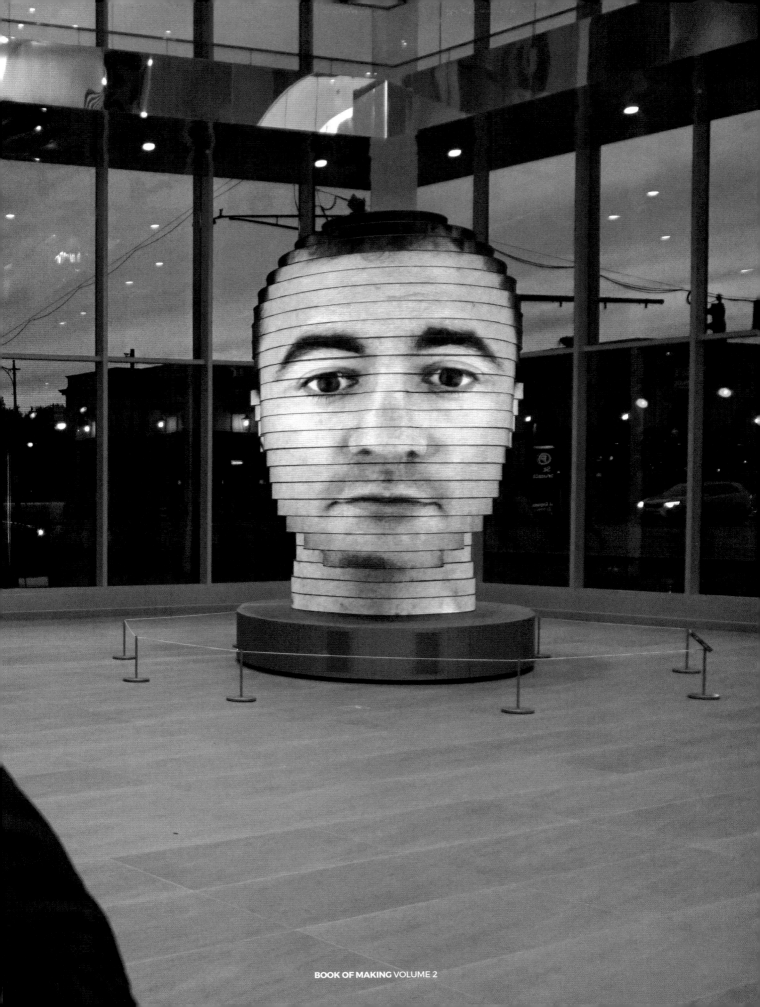

Right ↗
Each of the circular
mounts holds a
Raspberry Pi and
Camera Module

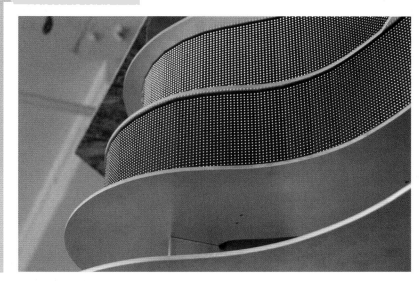

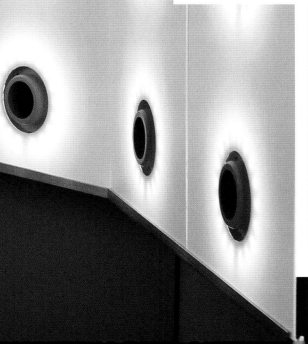

Above ◈
The frame is made from 24 layers of curved aluminium, stacked to form a human face

Left ◈
The eyes and nose of each subject are altered slightly in software to better conform to the physical sculpture

Below ◈
Inside the head

FEATURE

SPONGE

Scrub the blueprints for your builds with this versatile soaker

Mayank Sharma

🐦 @geekybodhi

Mayank is a Padawan maker with an irrational fear of drills. He likes to replicate electronic builds, and gets a kick out of hacking everyday objects creatively.

leaning tools are a modern-day necessity. They come in all shapes and sizes, but perhaps the most inconspicuous, yet indispensable, is the kitchen sponge. This cleanser can be traced back to the hygiene-conscious ancient Greeks, and has an interesting evolution that dates back to the beginning of life on the planet. The earliest Olympic participants bathed with handfuls of specially treated aquatic animals, known as sea sponges! The use of multicellular marine life as a cleaning tool was very popular through the Middle Ages, and has been documented in ancient Roman text as well. Sea sponges were thought to have therapeutic properties, which is why they were used by the physicians of the day to clean and treat injuries and wounds. Recent research by MIT has confirmed that sea sponges were, most likely, the first animals on Earth.

The sponge didn't enter the kitchen until the 1940s. Until then, everything from sand, soda, and pieces of rag were used to wash utensils. Sometime between the late 1930s and early 1940s, German chemist Dr. Otto Bayer managed to synthesise polyurethane foam in his lab. Following his lead, in their bid to improve polyurethane foam for commercial viability, some German scientists ended up with a defective batch that was full of air bubbles. While analysing the failed experiment, the scientists took a closer look at the spongy material and soon realised its potential, primarily because of its resemblance to the sea sponge.

It wasn't long before these polyurethane sponges were put to use as cleaning agents. But, like most discoveries, the earliest polyurethane sponges were very fragile and disintegrated easily when used in the kitchen for washing dishes. Also, while they produced good foam, they weren't suitable for heavily stained utensils, and would roll up and crumble quickly. The durability of the modern-day sponge was a result of refinement to its manufacturing process.

These days, you can find various types of sponges at the supermarket. Hand-sized cellulose sponges are the most common; inexpensive and providing good absorbency, they are suitable for all kinds of chores. Some cellulose sponges have an abrasive material on one side to help scrub patios, grills, and ovens. On the other end of the spectrum are nylon sponges that are meant for use on surfaces that scratch easily, like glass, porcelain, and kitchen counter-tops. They aren't as absorbent as cellulose sponges, but they can hold a good amount of liquid soap for effective cleaning. And, just as you can find sponges for all kinds of surfaces, our spirited makers have squeezed them for all kinds of hacks.

SPONGE BOT

P rolific maker and Instructables.com's Senior Community Manager, Randy Sarafan loves making robots from easily accessible knick-knacks. He's built the Sponge Bot using a couple of rulers as the robot's frame, a jar lid as its crank, and a paintbrush as the connecting rod. The robot was initially conceptualised as a rope-climbing robot: "I'd like to say there was some sort of grand inspiration for selecting sponges, but really it was just trial and error. Of all the things I have lying around, those

" A SINGLE ROTATING MOTOR CREATES A COMPLEX SCISSOR-LIKE BACK AND FORTH MOTION "

seemed to work the best. As I often tell my sceptical wife, building robots isn't really an exact science," Randy opines. When he couldn't get the rope-climbing robot to climb, he tossed the lamp cord locking mechanism he'd designed to help it climb, and replaced it with a swivel ball caster instead, and began attaching different things to the rulers. In the end, he ended up with a robot that uses a single rotating motor to create a complex scissor-like back and forth motion. As with all his Instructables, Randy has posted detailed build instructions that illustrate each and every step of the construction. □

Project Maker

RANDY SARAFAN

Project Link
hsmag.cc/eFAibc

Above ◈
If you haven't built a robot yet, take Randy's free online Robot Class (hsmag.cc/oaznBO) to get started

FEATURE

PUNCHABLE KEYBOARD BUTTON

Project Maker
AMAL MATHEW

Project Link
hsmag.cc/epJCNj

Right ◈
A Facebook video on how to deal with stress at work, which featured a punchable keyboard, inspired Amal to hack one of his own

*I*f you've worked with computers for a long time, there have probably been several times when you've felt like punching the keyboard in frustration. Amal has put a sponge's elasticity to good use to help you vent your emotions without any monetary loss. He's used a piezo element, a high-voltage 1 megohm resistor, and an Arduino-based ATmega32U4 board, with a built-in USB that helps identify it as a keyboard. When the piezo element records the voltage generated by a punch, it's read through the analogue pin of the microcontroller,

which translates it into a key press. "I used two sponges and placed the circuit in-between it," shares Amal, whose mother helped stitch a cloth enclosure for the venting device. Follow his detailed instructions to build and code the circuit. Just enter the ASCII value of the key you want pressed when you punch the keyboard, and you're all set. Besides being an outlet for your emotions, since the punchable button is completely customisable, Amal suggests you can use the device as a short cut for entering a complex password, or for some other action you perform regularly. □

SPONGE MUSIC

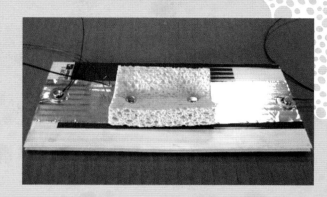

Keith is the Director of Learning Technologies Center, at the Science Museum of Minnesota. A Scratch veteran, who has tinkered with the block-based visual programming language for over a decade, he came up with the Sponge Music project as a means to demonstrate variable resistance using Scratch. Here, he used a PicoBoard that has a bunch of sensors, such as a light sensor and a sound sensor, and can interface with Scratch. The Instructables page has an illustrated guide to help you build the music-oozing sponge sensor. You'll also need to download Scratch, and import Keith's Scratch project (**hsmag.cc/CsQMxt**). Depending on how much force you exert to squish the sponge, the contraption will pass electricity from one connector to the other that turns it into a number, thanks to the PicoBoard, which then plays the associated note as defined in Scratch. □

Project Maker
KEITH BRAAFLADT

Project Link
hsmag.cc/ySWoGd
—

Above ◈
Although we like to minimise the use of plastics, you'll need it for this build, since the sheet will constantly be wet, due to the sponges

SPROUTS ON A SPONGE

Project Maker
TAMMY DUBE

Project Link
hsmag.cc/luEpjR
—

Left ◈
If you think growing a sponge garden is an interesting idea, head to Tammy's website for all kinds of kid-friendly projects and experiments

Tammy was looking for a child-friendly activity to mark St Patrick's Day, when she came across an article on how to grow sprouts on a sponge. So, she set about cutting some pieces of green sponge into the shape of a shamrock. The cut sponges were then soaked in water to make them damp. She then sprinkled lettuce, spinach, and broccoli seeds onto the sponges, and pressed them into the holes in the sponge. The planted sponges were then placed on a windowsill, and her kids used a water sprayer to keep them moist. "We found it helpful to turn a clear plastic container over the plate at night, to keep the moisture in," writes Tammy. They had themselves a sprout garden in a week, and the kids were amazed by the fact that the seeds didn't require soil to sprout. □

SNES-style RetroPie build

Gaming like it's 1999

———————

By **Mark Thorpe**    hsmag.cc/UOlQuJ

At one end of the upcycling scale people are taking old furniture, repainting it, and adding new knobs/handles: it's simple, it makes an old thing more useful, and it works.

At the other end, there's this Raspberry Pi Zero W-based build from Mark Thorpe. It takes an old games cartridge and reuses it as the case for a Pi Zero W, running the RetroPie games emulation platform. Mark's taken something old, added something new, and retained the spirit of the cartridge's original use, so you get the Proustian rush of wasted 1990s afternoons when you pick it up. He's even added a Pimoroni LED SHIM to the display, for a bonus rainbow effect.

The materials for the build cost about £30 from your local internet retailer, and Mark has documented the build here: **hsmag.cc/UOlQuJ**. □

Above ◈
RetroPie gives you an interface that's easy to use with a controller

Right ◈
For the true retro experience, don't forget to blow into the cartridge before powering on

We Learn
WELDING

From beginners to slightly burnt beginners: learn from our mistakes as we take on a new skill

By **Ben Everard**

Hacking things isn't so much a skill as the intersection of a lot of skills. You may have to be able to design things, do some coding, solder bits together, and build something to hold it all together. The more skills you have, the wider the range of projects you can take on.

Personally, I have a strong background in computing. I can whip up code to solve most problems I encounter. Need a virtual machine set up and managed to handle the back end of a task? I'm your man. I can have a decent crack at digital electronics, and have a large stack of boards that I've soldered together over the years. However, when it comes to physically building stuff, my skills leave a bit to be desired. When pressed, I can assemble something out of wood that serves the purpose. When it comes to metalwork… well, let's just say I have to start from scratch. I'm always keen to find new ways of expanding my repertoire, so I set out to see how far I could get learning to weld in a day. To be more specific, can I learn enough welding to make a small stick man from a metal bar? Equipped with only rudimentary equipment (an AC welder and appropriate safety equipment) and a little instruction, I set about the task.

Electric 'stick' welding is, in principle, a simple thing. You connect one bit of metal to a power

Below
Our very first attempt to control the arc. You can tell from the welding splatter and occasional balling that we still need some practice

LEARNING TO WELD

Basic welding isn't hard, but it does require quite a bit of kit that can be a little expensive (expect to pay around £100 for a basic welding setup), and it's best used (at first) with some guidance from someone who knows what they're doing. Vocational schools often have short courses to help you learn the basics, or you might find a willing instructor in your local hack/hacker/maker space. Some vocational schools will let you use their equipment after you've gone through the training, which can be easier and cheaper than trying to set up on your own.

> A few sparks splutter out and the welding rod is stuck firm to the surface. A few wiggles pull it free,
> **but it's a fairly unimpressive start**

supply capable of creating a large current, and attach the other side of the power supply to a flux-covered rod known as an electrode. Hold an electrode close to the metal you want to join and a spark arcs between the two. This arc is hot enough to melt both the metal you're welding and the electrode. This all pools together to form a cohesive mass that forms the joint you're welding. The first step, then, is to learn to create this arc.

The beginner's approach to this is to run the electrode along the metal briefly, then just move it away slightly – an action not unlike striking a match. I have a plate of scrap metal clamped to the bench in front of me on which to practise striking an arc. The view through the welding mask is completely black at first, so I line up the electrode with the mask up. I flip the mask down and strike. Nothing. I try again. Nothing. A few sparks splutter out and the welding rod is stuck firm to the surface. A few wiggles pull it free, but it's a fairly unimpressive start.

HELLO WELD!

Frustrated, I run the electrode along the surface again. There are a few sparks, and this time, as I pull the electrode away, there's a dull yellow glow emanating from the gap between the rod and the metal plate. In my excitement, I pull away and the arc dies almost as soon as it was created. It's not a great arc, but it was definitely an arc. After a few more tries, I can create an arc, if not consistently, at least regularly.

Just as generations of programmers have started coding by getting the computer to utter 'Hello World', so novice welders often start by guiding the arc around in their name. This gets us used to not just holding an arc, but also manoeuvring it through the dull, almost black world that we view through the welding mask.

I line up, flick my mask down, and begin the B. The arc melts both the plate of metal the arc hits and the electrode. As the electrode melts, it's deposited on the plate and I leave a ridge shaped in the letters of my name. I flick my mask up as the tail of the N is still glowing slightly from the heat. Initially the letters are black and slightly crusty, but

WELDING VS SOLDERING

When thinking about welding, it's sometimes useful to compare it to something it's not, such as soldering. When you solder a joint, you heat up the things to be joined, then add a filler metal (the solder) which melts into the joint. When it cools down, the solder hardens and everything is held together.

The difference between this and welding is that at the end of it, there are still three distinct things: the two objects being joined and the solder. The joint holds together because it's all stuck together, but they're still different things and could be removed from one another.

When you weld, you melt all three things and mix the resultant pool of molten metal. As such, there's no clear line between one thing and the other, they just blend together. There's no such thing as unwelding (as there is with unsoldering). You can cut the joint, but you can't separate out the constituent parts.

OTHER TYPES OF WELDING

We tried to learn AC stick welding. Equipment-wise at least, this is the simplest form of welding. You just have a coil that converts 240 V (or whatever your local mains voltage is) to a lower voltage but a much higher current than is usually available through a socket (it can be well over 100 amps). The welding electrode is surrounded by flux that keeps the joint clean, but can also cause problems. It's a bit of a rough-and-ready type of welding that can be great for hackers but isn't always the best choice. Here are some other options:

DC – Stick

Similar to the AC welding that we did, but this time using DC electricity. The equipment needed is slightly more complex, but the general process is exactly the same as with AC.

MIG

Metal Inert Gas welding also uses an electric arc to melt the metal, but rather than flux, it uses an inert gas (such as argon, carbon dioxide or helium) to protect the weld, which can result in a neater join. The filler rod is automatically fed into the joint as you weld.

TIG

Tungsten Inert Gas welding also uses an inert gas to protect the joint, but unlike MIG welding, the filler rod isn't automatically fed in and it's up to the welder to apply this as and when it's needed. TIG welding is the most versatile form of welding, but it's also the slowest and most difficult to learn.

Spot

Very easy to do, but limited in what it can achieve. Spot welding uses two electrodes close together (typically on opposite sides of the joint) to create a point of heat as well as pressure to hold the objects in place. There's no filler rod, and the two objects are just melted and pressed together. This is usually used to join two flat surfaces.

Oxyacetylene

Unlike the others we've covered, this one uses gas to heat up the metal. A combination of oxygen and acetylene is burned in a welding torch which melts metal. A filler rod can then be used where necessary.

Brazing

This one isn't really welding, but more like high-strength soldering: you heat up the bits of metal you want to join and use this heat to melt a brazing rod into the joint.

with a metal brush removes the burnt flux, leaving just the letters raised up in metal.

That's not really welding, it's just depositing metal, but I feel hugely satisfied by the feeling that metal – which in my imagination is solid and immutable – has bent to my will.

Stage one of learning to weld is done, it's now time to join some bits of metal together.

In principle, this is simple. Using exactly the same technique I used to write my name on the metal, I need to melt both sides of the joint and deposit a little metal into the gap between them. If all goes to plan, this should cool and solidify to a single piece of metal with the three parts mixed together in the middle.

LET'S BUILD SOMETHING

Mask down, arc struck, I begin to move down the joint. Immediately I see where the skill in welding comes in. The instant the arc is established, the metal starts to melt. Move it too soon and nothing's melted enough to stick together. Leave it too long and it's too melted, leaving a hole in the metal. I need to carefully glide it down, gently moving the electrode between the two surfaces to be joined as it travels down the seam. Of course, I fail spectacularly at that. I move in jerky motions, and try to cheat by moving back up the seam to a part where I'd moved too fast.

The welding rods are coated in flux that protects the joint from oxidisation (it serves a similar purpose to the gas in MIG and TIG welding – see

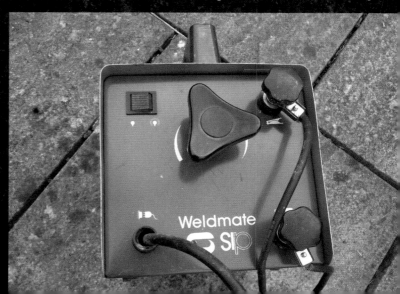

Below ◈
My equipment for the day was an aging Weldmate that used to belong to my grandfather. Basic equipment is fine for basic welding

Right ⊿
The finished product.
Metal sculptors may
not be fearing the
competition just yet,
but I'm proud of it

boxout). This should melt and rise to the surface of the joint; however, if you move back up the weld, this flux gets embedded in the joint and you get a stick-shaped hole.

Ugly welding is acceptable, though. The question is, how strong is it? Time to give it a bend and see if I can break it.

Arrghnnn!

The freshly welded metal holds quite a bit of heat and even through hefty gloves, it singes my fingers. Only slightly though. After waiting another minute for it to cool down, I try to snap the joint. My first joint is ugly, but surprisingly strong. Despite it

looking like someone tried to nail-gun two bits of jelly together, I can't break the joint with my hands, so I'm chalking that up as a win.

We wanted to find out if you could learn to weld in a day, and for that we needed a test. For us, it was whether or not we could weld together a simple stick-man sculpture (Maybe sculpture is a little too grand a term, but you get the idea).

The welding here is harder as it's at strange angles, and the clamps holding things in place can get in the way a little, but it's all the same basic process: strike an arc, position the arc in the seam, and move the arc to blend the metal together to create a solid joint. Some are, ahem, more successful than others, but the end result, after only a few hours from the first time I held a welding rod, is a solid structure.

The essence of hacking is expanding your skills range, and welding is a great area to move into. It doesn't take too much time to grasp the basics and even basic skills can be useful. I can't claim to be a competent welder, but I do feel that I now have a new skill that I can bring to bear on things I make. It might be a while before I'm doing anything load-bearing or which has to look good, but a custom robot frame or a jig to hold things in place is now within my repertoire – just. ⊡

HACK YOUR OWN WELDER

In essence, an AC arc welder is just something that can supply a lot of current (generally a minimum of 50 A). There's nothing particularly complicated in this, and we've seen plenty of home-made welding setups, both by using coils to increase the current in mains power or by amalgamating enough batteries to supply current.

It's a fascinating project if (and this is a big if) you have the skill and experience to do it safely. Remember that you need to create enough current to melt steel. That means that there's enough current to do a lot of damage to just about anything that gets in its way, including a human.

One thing that you should never attempt to make yourself is goggles. Eye damage in welding comes from ultraviolet light, so it can be hard to know if protection is adequate until after any damage has been done. Look after your eyes – invest in good quality goggles.

Illuminated waterways map

By AlexT9   🢅 hsmag.cc/WwpRCq

" **had just discovered my local makerspace, and was
immediately drawn to the laser cutters.** I have always
enjoyed looking at unique maps, so I knew I wanted to do
a map-related project. At some point, I got the idea to fill in
laser-etched wood with opaque-coloured epoxy. The idea
grew and became way more complex (as most projects
always do…) into fully cutting through the wood and backlighting
the epoxy. I used Photoshop and Illustrator to convert a high-
resolution waterways map into the vector format needed by the
laser cutter. After cutting the wood, I used a blue pigment powder
with two-part epoxy resin to fill in the waterways. Standard LED
strip-lighting provides the backlight for the resin. I think the coolest
aspect of the map is being able to easily visualise different river
basins. The Mississippi river basin is huge!" ◻

Right 🢒
It took Alex about
eight hours to find
the right level of
detail for the rivers

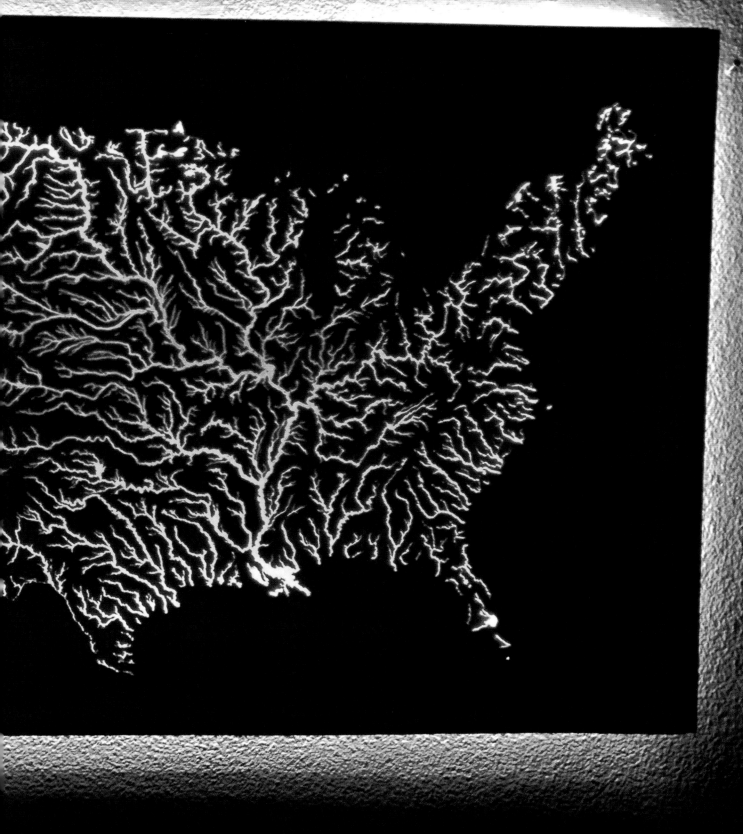

REVIEWS

HACK | MAKE | BUILD | CREATE

Expert reviews of some of the best products and components available for makers

PG
174

ELEKSDRAW

EleksMaker's pen plotter put to the test

REVIEW ━━━━━━━━━━━━━━━━━━

Neje DK-BL laser engraver

Can a cheap laser cutter cut it?

NEJE ◈ $77.28 | trusfer.com

───────

By Jo Hinchliffe 🐦 @concreted0g

Below ◙
The tiny Neje DK-BL engraver. We're not going to engrave any skateboards or coffee tables in it!

Laser cutters are great bits of kit, but they can be big and expensive. They have to be, because powerful lasers are, well, big and expensive. However, if you want to engrave rather than cut, you can get by with a smaller laser, and there are a variety of cheap laser engravers available.

We put the Neje DK-BL through its paces to see how well it works. This engraver is Bluetooth-enabled, contains a rechargeable battery for portable operation, and sports a 1500 mW 405 nm laser. It arrived well-packaged, with a collection of additional bits, a USB lead, some material samples, a small paper manual, and a CD with some software. The manual also provides a QR code that links to an application for either Android or iOS.

The moving innards of the engraver are small and (as discussed a lot online) they appear to be manufactured using leftover stepper motors and assemblies from optical drives. On powering up, the X and Y axes move to their limits and then reset to the centre in a slightly unusual way. They have no limit switches that signal the end of the axis travel, so instead the machine plunges the axis towards each end and it grinds at the limit momentarily before returning a known amount to centre. This works, but it is hugely inelegant. We question how well this will work if used every day for a year, but it worked for us.

SET LASERS TO ENGRAVE

Having stuck a workpiece under the rubber bands, to hold it to the table, we need to focus the laser to the smallest point possible. This is done by a focusing ring on the laser module itself. Whilst the thread on the focusing ring is crude, it is straightforward to focus on a variety of material heights. The machine has a very small working area of 42×42 mm but it seems, in experiments, to be able to accommodate taller materials, and we have successfully engraved into a 25 mm diameter tube, and the documentation reports it can work on objects up to 78 mm high.

Turning on Bluetooth on our phone and launching the app, the Neje paired successfully and appeared ready for work. The app is typical of these cheaper types of machines and has numerous faults. For example, when browsing for an image file, if you select back beyond the home location of the file browser, it crashes. The application also has numerous functions that make little sense; that said, after a few experiments we got to grips with the app and were able to create repeatable engravings

> **Prudence suggests that it's better to err on the side of caution, and pair this with safety glasses**

onto objects. It has simple features to allow you to write and place text and convert images to engrave. There's also a very handy feature where you can set the laser to show you a preview outline of where it is going to burn the engraving, and you can jog this rectangle around live until you are satisfied you have it in the right place.

DESKTOP DESIGNER

The Windows application is a similar story. Like the Android app, it works and allows you to set the engraving burn time of the laser. When using the PC software, the engraver is connected via USB and you get some information about the charge level of the internal battery (which is handy as it didn't seem to indicate charge level when plugged into a charger). Our main gripe with the Windows software is that it seems inconsistent and often reluctant to connect to the machine. When it does connect, it worked OK, but during testing on both a Windows 7 and Windows 10 machine, it took numerous attempts at plugging in and turning on to get the machine to connect to the laptop.

The Neje has a small piece of tinted plastic which connects to the front of the machine, with magnets to provide a filter to protect the user from the laser. The sides and rear of the machine are open and so it's possible to view the laser there and damage your eyesight. Prudence suggests that it's

better to err on the side of caution, and pair this with safety glasses for the laser wavelength.

However, all this said, the results are good from the machine. This isn't going to cut through any material thicker than paper (which it can certainly do well) but, if used for its main purpose of engraving, it works well across a variety of materials with high accuracy. We tested across varied card stocks, plywood, solid timber, and Perspex and it engraves well and yields great results.

Left ◈
A selection of items engraved with the machine in a variety of materials

REVIEW

Sony Spresense

Hexa-core, GPS-enabled, AI-ready microcontroller

SONY ◆ From £49.19 | sony-semicon.co.jp

By **Ben Everard**   🐦 @ben_everard

There are three Spresense parts from Sony. A base board (£49.19) gives you a system on a chip (with GPS built-in), 17 digital GPIOs, two analogue inputs, and four LEDs. You can plug this into the expansion board (£32.39) which then gives you an additional twelve digital I/O pins, four analogue inputs, and a headphones connector. There's also a camera (£35.88) that slots into a connector on the base board.

The brains behind the base board is a six-core Cortex-M4F processor running at 156MHz. For anyone familiar with ARM processors, this will sound like a microcontroller – the M-series is found in boards such as the Adafruit Grand Central and Arduino Due, whereas the A-series is the type of ARM core found in more general-purpose computers such as the Raspberry Pi.

A BIT OF THIS, A BIT OF THAT

This board, however, with its six cores has quite a lot of resources to manage, and sits somewhere between the two. Although it can be programmed via the Arduino IDE, it runs a stripped-down UNIX-compatible operating system called NuttX. If you program the Spresense using the Arduino IDE, you may never know that there's this OS behind the scenes managing things, as it works just as any other Arduino-compatible board does. However, you can delve right down into its internals using the Spresense SDK. You can even connect to the UNIX shell-like command-line interface over a serial connection, and interact with the board this way. Development via the Spresense SDK officially requires Linux; however, we were able to get it working on Windows 10 using the Windows Subsystem for Linux. It's not a particularly straightforward process, though, and probably not worth it for casual users. It does, however, let you unlock more performance due to the multi-core architecture. Although the Spresense has six cores, only one will run your application. The remaining five can be used to run specific, additional tasks. Sony provides

Right ⬈
All three parts connected together just waiting to be turned into a computer vision-enabled, GPS-tracking quadcopter

some code for running audio code on these cores from within your Arduino sketches. You can find more details of these at: **hsmag.cc/luPaPI**.

Sony supplies a set of Arduino libraries to help you get the best out of this board in a few areas, as well as audio. The DNNRT library allows you to use deep neural networks (DNN) created using the Neural Network Console – a training tool that lets you configure and train your neural networks on your main computer before transferring this trained brain to your Spresense. While this isn't completely straightforward, it doesn't require a lot of programming skill, just the knowledge of how to configure a network.

BONUS HARDWARE

The module has built-in satellite positioning that uses GPS, GLONASS, and QZSS satellite positioning systems. We found that the antenna wasn't particularly sensitive and failed to get a fix inside at a location our phone could get a fix at. That said, you're unlikely to be relying on satellite positioning inside.

The base board includes a camera connector, and Sony has released an official camera module. By microcontroller standards, this camera – at 5 megapixels – is really high resolution. While you might be used to cameras with far more pixels on your phone or Raspberry Pi, these cameras are rarely compatible with microcontrollers and the high data rate needed to operate such a camera. When combined with the AI capabilities on this board, this has the potential to add some really powerful vision capabilities to your makes.

While the Spresense is a very capable board, it's not a beginner board – its unusual hardware means that it works a bit differently to most microcontrollers. The libraries that unlock the power of this board are well-documented and do make it reasonably straightforward to use this particular hardware, but it is a bit different from other Arduino-compatible boards.

There are a few oddly disparate areas where the Spresense can really stand out. The ability to play or record sound in the background – without taking up processing time on the main CPU – could be a huge boon. The hi-res camera can be really useful,

Above ◈
The base board itself packs a huge amount of power into a very small footprint

as can the AI capabilities. Looking at these, they read like a set of scenarios more suited to small Linux-capable computers, such as the Raspberry Pi, than microcontrollers. Perhaps this isn't surprising, since the Spresense runs a UNIX-like OS – though one that much more stripped down than Linux.

The Spresense is more powerful than most microcontrollers (particularly when the extra cores are taken into consideration), yet it boots in under a second and runs real-time code, so can be used for precise hardware control. While it's not for every project, the particular capabilities of the Spresense make it an excellent addition to the arsenal of hardware available to makers. ◻

VERDICT

An unusual microcontroller with great potential for AI and vision projects.

9 /10

Mayku FormBox

Vacuum forming for your desktop

MAYKU ◈ **£599** | Mayku.me

By **Ben Everard** 🐦 @ben_everard

This vacuum former comes without one thing – a vacuum. You need to plug in a regular household vacuum cleaner and use this to suck the softened plastic against the mould. There's a socket on the Mayku that you plug the vacuum in and this turns on the vacuum at the right time. This does require a vacuum that turns on when the socket is turned on, and it also requires a vacuum that can slot into the rubber adapter. Our venerable old Henry worked fine. The maximum power the Mayku can handle is 2000 W, but most vacuums should be below this. Our Hoover defaults to a weak suction unless a high-power button is pressed, and we had no problem using it on low-power mode, so wouldn't expect a problem with other machines.

To vacuum form something, you need a mould to form it around. The mould goes on the bed, and a sheet of plastic is clamped into the frame. The sheet is softened by the heating element, and plunged over the mould. At this point the vacuum comes on and sucks the plastic over the mould. It quickly cools and rehardens in the shape of the mould.

MATERIAL SELECTION

The 20 cm by 20 cm work area isn't huge, but keeping things compact means that the device fits into even small workshops or offices. You can use a wide range of thermoplastics in a vacuum former, including PETG (which can be food-safe), ABS, and polystyrene. Mayku itself sells two types of sheets: Mayku Cast (0.5 mm thick, food-safe, and clear plastic), and Mayku

Right ◈
The concept is simple: a heat source, a sliding holder, and a suction bed combine to help you manufacture items quickly and easily

Left ◈
Our vacuum forming experiments with household objects produced great results

Form (0.5 mm thick, white plastic). Unsurprisingly, Mayku Cast is for creating moulds and Mayku Form is for, well, forming things. Both are available in packs of 30 for $39.99.

Vacuum forming is an analogue technique – you form with a physical mould, rather than a digital design. You can carve the design out of wood, or other materials – we even found potatoes worked. Alternatively, you can use some object you already have as a starting point – we tried Christmas decorations. If you'd rather use digital tools, you can 3D-print your mould – vacuum forming will allow you to duplicate this far faster than 3D-printing each object.

All the fittings are quite solid. Moving the trays requires a firm, decisive movement which inspires confidence in the build quality, but also can give quite a jolt.

TIMING TROUBLES

The FormBox has a timer that should allow you to heat a sheet the correct amount automatically. However, we found this to be essentially useless. We were unable to dial in temperatures that were accurate enough to be useful, and we do our mouldings by eye. Both the standard materials – Mayku Form and Mayku Cast – noticeably deform once they're ready to go, so are easy to use without this timing. With other materials, it might be more useful to use the timer.

Vacuum forming is still relatively rare in the maker community – far more makers have access to laser cutters and 3D printers. This means that there's a little less knowledge in the community for this style of manufacture than others. This gives you a chance to wow your fellow makers with your vacuum forming skills.

> **"** It was about an hour between first opening the package and **having our first bits made "**

What vacuum forming can make, it makes well and quickly. A minute or so to melt the sheet, then a few seconds to suck over the mould and you're done. For short-run manufacturing, you can churn out your builds far faster than most other desktop fabrication techniques. What really impressed us was how quickly we could get started. It was about an hour between opening the package and having our first bits made (a phone hologram projector that comes as a sample mould). It would have been quicker, but we failed to realise that the bottom seal moved up (as we said, the moving parts lock together quite firmly). Five minutes later we were creating things we'd designed ourselves. This is a tiny fraction of the time it takes to get started on any other desktop fabrication tool.

However, as with all manufacturing techniques, vacuum forming does have limitations – you can only make shells or casting moulds, and even then, only if there are no undercuts. Still, there are plenty of cases where the advantages outweigh the disadvantages (such as if you need to recreate your designs quickly, or if you're running a workshop where you want to help people make something quickly).

Vacuum forming complements more common maker manufacturing techniques such as 3D printing and laser cutting by being good in areas they're not, and when it comes to vacuum forming, the Mayku is the best available option for home or small workshop manufacturing. ◻

VERDICT
Easy to use and a valuable addition to desktop manufacturing.

9/10

Blimpduino 2

Taking to the skies – the sedate option

JJROBOTS ◆ From £75 | jjrobots.com

By **Ben Everard**   @ben_everard

There are lots of ways to get your projects airborne – we've looked at two options in past cover features: drones and rockets. However, there's a less common, more calm, and controlled option: blimps. These use a lighter-than-air gas (typically helium) in a balloon to lift the weight, then use several small motors driving fans to manoeuvre the aircraft around. Since there's no weight to lift (the blimp should be neutrally buoyant so neither fall nor ascend when floating free in the air), these motors don't have to be particularly powerful, and so a blimp needs much less power than most aircraft. The

downside of this setup is that it's much slower and much more susceptible to wind.

The Blimpduino 2 is, as the name suggests, an Arduino-compatible blimp. All the electronic functions are packed onto one board, which has an ARM Cortex-M0, accelerometer, LiPo charging port, and lidar. Additionally, the kit comes with three motors with propellers. You can also order a 3D-printed gondola ($9.12) if you don't have a 3D printer, a LiPo battery ($7.81), and a helium balloon (from $5). The only thing JJRobots doesn't supply is the helium, which you should be able to source from local party shops.

JJRobots recommends that its blimp be flown inside. Although the hardware should continue to work upwards for quite some distance (until the air pressure changes enough to stop the helium lifting enough weight), the lidar distance sensor is only accurate to four metres.

Although the blimp is Arduino-compatible, this isn't actually relevant for the basic operation. Once the hardware's connected up and powered on, you can control it via a phone app. This gives you basic controls for flying. We found that the app didn't work properly on the first phone we tested (there's a thread about this on the forums – **hsmag.cc/uZJmbn** – but as we go to print, this hasn't made it into the published app). However, we were able to get it working on another phone. This gives you a basic, WiFi-controlled remote. You can steer the blimp, and take it up and down. These controls aren't simply connected to the motor output: there is enough intelligence in the Blimpduino 2 to compare your instructions to the received data from the accelerometer and lidar so it will behave intelligently, not just spin out of control.

Below ◆
The assembled control structure, ready to be attached to a balloon

'Attempt' is an important word here as it's limited by the hardware as to how effective it will be. For example, if your balloon is poorly balanced, it might not be able to ascend and descend properly.

It's a little hard to give good estimates of how long it will fly for, as the amount of energy it has to expend depends so much on how you fly it. In ideal conditions, it should be able to hold its altitude without expending any energy, so could hover indefinitely. However, you're probably going to be

> **In ideal conditions, it should be able to hold its altitude without expending any energy**

moving it around. We found we got about 20–30 minutes from a 500 mAh battery between recharges, but you may find you get different results.

LOOKING INSIDE

If you want to go further, the firmware behind the device is written in Arduino C++, and the programming port is open, so you can add whatever extra features you like, whether that's some form of autopilot or additional sensors. You can add more weight without it being a burden on the motors, provided you've got enough lift capacity in your balloons. You can add larger (more) balloons, but this will eventually slow down your blimp and make it harder to control.

While there aren't any specific guides to modifying the hardware or software, the source code (available here – **hsmag.cc/SHSBAu**) is commented and reasonably straightforward to

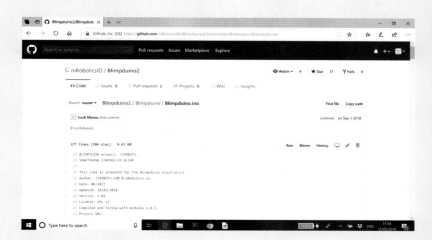

follow. Perhaps the most unusual part of it is that the default app communicates via UDP, rather than the more common TCP. This isn't particularly hard to use, but it's less likely to be familiar to users. There is no need to stick with this communication method – you've got full control of the hardware, and can communicate with it however you like. The only slight issue is that you don't have access to the programming port on the ESP8266 used for WiFi, so you'll need to be familiar with the commands for this if you want to modify the wireless setup.

We really enjoyed how adaptable the Blimpduino 2 is. In its most basic use, it's a remote-controlled blimp (some assembly required). You can switch out the balloon and 3D-print replacement parts if you need, and you don't need any particular technical skills to get to this level. However, if you like to tinker with things, it's easy to dive further in, and everything's accessible to you. You can add other things that don't require programming, such as an off-the-shelf FPV camera/transmitter module (these just need power and will transmit automatically without programming). If you do want to go further, there is the hardware to support it. □

Above ◈
Go forth and code your own blimp – take and modify the code from hsmag.cc/cvIMBZ

Below ◩
The full pack – just add helium!

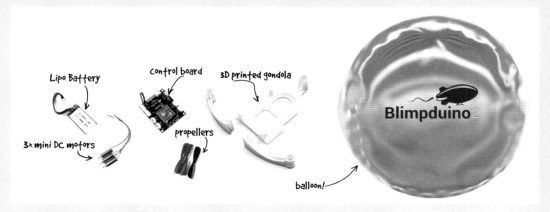

Lipo Battery
Control board
3D printed gondola
propellers
3× mini DC motors
balloon!
Blimpduino

VERDICT
A fun flying machine, suitable for all abilities.

8/10

REVIEW ━━━━━━━━━━━━━━━━━━━━━━━━━━━━━━━━━━

PyPortal

A one-piece internet counter for almost any internet data

ADAFRUIT ◈ From $54.95 ⏐ adafruit.com

By **Ben Everard**   🐦 @ben_everard

The PyPortal combines a 320×240 colour TFT display with a Cortex-M4 processor and an ESP32 WiFi module to create a web-based display programmable in CircuitPython or Arduino.

The most basic use of the PyPortal is as an information display. The CircuitPython library includes the ability to do this almost automatically. Feed in your WiFi credentials, the URL of the data (in JSON format), and the bit of data you want to extract. Pair this with a background image and the location on the screen you want the text to appear, and you've got your Internet of Things display. Adafruit has guides using this basic formula to display stats for social media (such as YouTube subscriber counts or Reddit viewership), environmental data such as air quality, and text via a quote of the day.

Of course, you don't have to use the PyPortal like this. It's completely customisable – and compatible with the displayio CircuitPython library, which allows you to easily place text and graphics at different points on the screen. At the moment, performance isn't great – of the order of two or three frames per second, so animations aren't going to be smooth, but perfectly fine for static images and slowly changing text. The hardware is capable of much more than this, and it's currently slowed down by the CircuitPython libraries (which should hopefully see a speed boost in future versions). If you need animations, there's Arduino support, where we were able to get >20 fps on animations (take a look at the graphics demos here for examples: **hsmag.cc/jHheml**).

If your main exposure to touchscreens is using phones, you need to be aware of the technology here. It's a 320×240 pixel 3.2-inch display with resistive touch. This has 125 pixels per inch (PPI). The iPhone X has 458 PPI,

Below ◈
The PyPortal is only slightly bigger than the screen, so it's easy to embed in a picture frame or other mount (there are several 3D-printable designs available from the Adafruit website)

and other modern phones are similar – of course, they're also a lot more expensive and a lot less hackable. For text and simple graphics, the PyPortal works well. For photographs and highly detailed images, you're probably going to struggle. Similarly, the resistive touchscreen on the PyPortal works well for simple finger taps, but doesn't have the sensitivity of high-end capacitive touch displays.

A built-in speaker and connections make it easy to give your projects audio, as well as visual, output. There's also an on-board temperature sensor (though the display can warm the whole board up a little), and an ambient light sensor which could be used to change the brightness of the display based on the conditions.

You can interact with extra hardware via an I²C bus (5 V, but can be converted to 3 V with a cuttable trace) and two digital IO/analogue-in pins, all of which are in Grove-compatible connectors.

USER EXPERIENCE

For a web-based data display, the out-of-the-box experience with PyPortal is brilliant. Grab one of the examples (such as the Reddit subscriber count here: **hsmag.cc/kToalq**), change the JSON source, the background image and caption, and you've got an internet counter. The range of examples has grown significantly between when we started reviewing the PyPortal and when we went to press. If you're already moderately familiar with Python and JSON, you can probably get something working in under half an hour.

Going further in CircuitPython can be a bit tricky at the moment. The displayio module is new in CircuitPython 4, and the documentation is still catching up. That said, there are some examples out there that show how to use it – take a look at the weather display example, particularly the **openweather_graphics.py** file or the HalloWing Magic 9 Ball example (**hsmag.cc/SISvKr**), for details. The basic idea is that the displayio root contains a list of nodes. Each node can be either a thing to be drawn or another list. These things are then drawn in the order they are in the list (so things go over or under each other in this order). You can add text or graphics in this way, and manipulating them automatically updates the display. Of course, things are improving all the time, and when you read this, there may be more information available.

Using Arduino, the path is a bit more well-trodden. The ILI9341 chip that drives the display has a well-tested Arduino library that works with the classic GFX library.

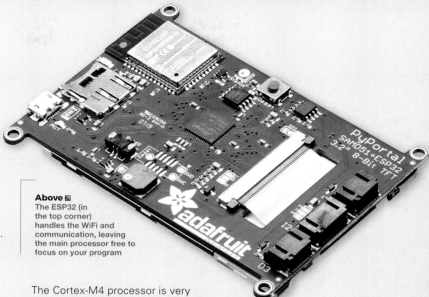

Above ↗
The ESP32 (in the top corner) handles the WiFi and communication, leaving the main processor free to focus on your program

The Cortex-M4 processor is very heavyweight for such a product, especially as the data connection is handled by the ESP32. CircuitPython is getting faster all the time, but it's still a bit of a resource hog, so having this power gives you the space to do a reasonable amount of processing. If you use Arduino, rather than CircuitPython, then you've really got a lot of processing power to play with.

You do have access to the programming pins on the ESP32, so in principle you could do more with the processor it has if you want to, whether this is more processing of the data flowing in and out, or activating some of its other features such as Bluetooth.

Overall, the PyPortal is a great one-piece solution to displaying internet data. It's got powerful hardware, is easily expandable, easy to program, and is great value for money. ☐

Below ◈
The weather station example code lets you know what's going on outside without having to open the curtains

VERDICT

The best hackable out-of-the-box IoT display available at the moment.

9 /10

REVIEW

EleksDraw

EleksMaker's pen plotter put to the test

ELEKSMAKER ◆ £115–£140 | eleksmaker.com

By **Jo Hinchliffe** 🐦 @concreted0g

If you're ever wondering whether you need a plotter, head to Twitter and take a look at the hashtag #plottertwitter. You'll discover a community creating amazing images on both vintage pen plotters and modern machines.

The EleksDraw, by EleksMaker, is a kit which sits firmly at the budget end of pen plotters (around £115–£140 currently online) and consists of two sets of rails, to form an X and Y axis, with a belt system to move around a pen affixed to a lift mechanism.

The EleksDraw arrives extremely well packaged, and with all parts to complete present. A few extra of each type of bolt and nut are included, which is welcome for the 'I've dropped one and it has disappeared' scenario. Inside the box there is a small printed card, with a link to the EleksMaker Wiki, where the build instructions are to be found. The instructions consist of a series of pictures of each build step, and each picture is annotated with how many nuts and bolts are used in that step, and which

size they are. A little tip is to have a ruler present to double-check you have the right length component. For example, there are some 8 mm standoffs and some 10 mm ones, and they are easy to confuse without measuring.

We found we had to go back a few times as we put things together incorrectly. It took us around three hours to build, and there were a few areas we found more tricky, and a couple of problems we had to research. The EleksDraw forum (**hsmag.cc/hdtDcK**) has advice from other people who have been through the same process.

> **Connecting the electronics and motors** was an easy plug-and-play affair

Getting the belt tension on this machine takes a little time. As the belt is a toothed GT2 type, it's easy for it to leave some slack in just one section of the belt, and you only realise when you move the axis by hand. The tension needed to be tighter than we first imagined but, once it was correct, the clamp that traps the belt on the rear of the machine locks it well, and it has required no further adjustment in use.

MOTORING ALONG

Connecting the electronics and motors was an easy plug-and-play affair. The next step was to download and install the software and drivers. The link to a folder of Windows software is on the instruction Wiki page. After installing the driver, instructions are given to select which of the EleksDraw machines you have, as the 'EleksCam' software is built to control

Below ◆
Once assembled and a few tweaks are ironed out, the EleksDraw is capable of very fine and accurate work

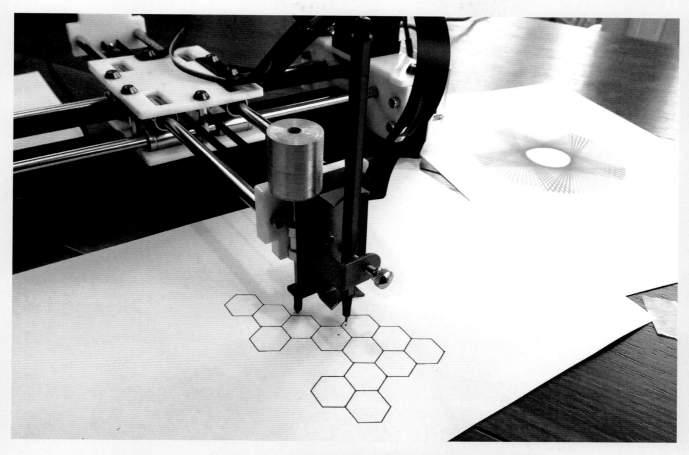

the whole range of EleksMaker products. We then turn on the machine and are nearly ready to draw something. The software is reasonably intuitive. There is a 'jog' control, allowing you to move the pen plotting head and a set home button, so you can zero the machine when it's in the desired start position.

One thing to note is that this machine has no end stop switches, so it is possible to crash the machine into its endpoints. As it is belt-driven, the belt will eventually slip, meaning that it doesn't damage itself much in a crash, but it's worth being aware of as it will ruin the image it is working on. To check the area the plotter will operate in, there is a preview button which moves the head, with the pen lifted, to show the area it will cover for the design currently loaded. Pressing the 'Laser ON/OFF' button in the software should lift and release the pen-lifting assembly, but this revealed that the assembly was far too tight to actually fall under its own weight. We stripped it down and rebuilt the assembly many times, trying to make it as free as possible. In the end, we used some fine wire wool to polish the metal slide rods, using some fine oil to eventually get them free enough. It's also possible to hang a couple of nuts or washers onto the assembly to add a little weight to help push it back down to the page when the servo releases it.

The EleksDraw will deal with a range of file formats, including SVG, which is a great vector format widely used by Inkscape users. We've thrown a range of files and vectors at this machine and it is really good and accurate and capable of brilliant work. We have found, however, that it erroneously scales vector images – we're exploring workarounds and solutions in the community and discussion forums. It's also an addictive machine: there is something brilliant and compelling about watching a pen draw an image under robot control! ◻

Above ◈
The EleksMaker EleksDraw, having plotted some generative vector images with a fine-tipped felt pen

VERDICT

Pen assembly frustrating to get correct, software and instructions a little vague, but a great machine capable of producing really high-quality work.

8/10

REVIEW

SparkFun Deluxe Tool Kit

A workshop in a box

SPARKFUN ◈ **$224.95** | **sparkfun.com**

By **Ben Everard**   🐦 @ben_everard

If you're kitting out a workshop for electronics, what do you need? Well, a soldering iron – that's a must of course – then there's tweezers and diagonal cutters. Oh, a pad can be useful. There's that thing of course, you know the one, it does the gripping in different angles – what's it called? Are we forgetting anything? Wouldn't it be useful if someone just put everything together into one place? That's exactly what SparkFun has done with its two tool-kits: the Beginner's Tool Kit ($49.95) and the Deluxe Tool Kit ($224.95). We took a look at the Deluxe version, which includes a Weller soldering station, heat gun, multimeter, and 20 other little bits and pieces that you need to make great electronics.

Keeping a kit like this affordable, while complete, is a balancing act. The hardware needs to be good enough to provide users with a pleasant experience, but at the same time, keep the cost manageable. Everything in this kit works and is perfectly good enough to get started with, but there's not many bells and whistles, and at this price, that's what we'd expect – after all, you could spend more than the price of this kit on just a multimeter. Looking at this multimeter, for example, it has all the basic features that you need for electronics, but if you're using it regularly, getting an auto-ranging meter can be useful. The soldering iron is a decent iron with adjustable power, but for trickier jobs, better temperature control may be

Right ◈
Two spools of solder wick are included because we all make mistakes a little more frequently than we'd like

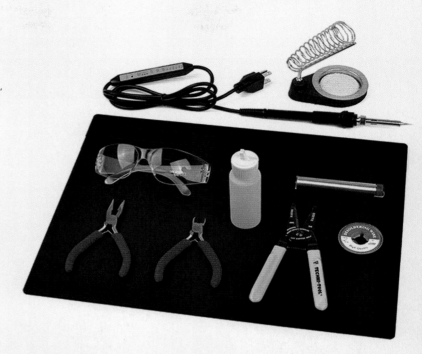

more useful. Conversely, the third hand, for instance, is much nicer than the cheaper options, with hinged metal joints.

MAKING COMPROMISES

Knowing where to draw the line on price and functionality is a difficult skill. This kit has clearly been put together by people who are (or have been) hobbyists, and know what you need. You could get this and never need another bit of kit for building electronics, other than perhaps replacing the consumables like wire and solder – well, the wire anyway, the 113 g (or quarter of a pound if you prefer) of special-blend lead-free solder should last you quite a while.

There are a few big-ticket items that more advanced hobbyists might use that aren't included (such as a bench power supply and an oscilloscope), but these aren't needed for getting started, and would add quite a lot of cost to the kit. These are also things that we got quite far without having, and certainly aren't essential to hobby electronics.

The big advantage of buying things in a kit is that you get the bits you need even if you don't know they're the bits you need, and this kit really delivers in this area. It's nice to see little things like the screwdriver or craft knife that could easily be missed out of a soldering kit, but at the same time, they're things you need. Sure, you might have them at home, but if you're kitting out a workshop, having them all together in one package is ideal. There's even a small water bottle included, for keeping water to top up your soldering iron cleaning pad. Another slightly surprising, but useful, entry is the heat gun. Again, this isn't something you'll typically see in electronics kits, but is actually a useful bit of hardware for a couple of tasks: heat-shrink tubing, and hot-air rework. The former of these is basically an upgrade to electrical tape for keeping your connections insulated, and the latter helps for fixing mistakes or breaks in PCBs, particularly with surface-mount components (it's not a full-on hot-air rework station, but can just about be pressed into service for simple jobs).

HIGH VOLTAGE

People outside of the USA, be warned, the soldering iron and heat gun are only rated to 110 V, so you'll need a transformer, as well as a socket adapter, if using it somewhere with different voltages.

You can buy most of the parts from SparkFun separately (only the water bottle isn't available), but

it is (by our count) about 10% cheaper to get the kit rather than buying the individual components, so even if there are one or two bits you don't need, it may still be cheaper to get the kit and then lend these to a friend in need.

This kit only contains the bits for putting together electronics, and not the bits for electronics themselves. There are no resistors, Arduinos, cables, breadboards, or anything like that. This kit

Above ◨
The Beginner's Tool Kit has everything you need to get you started

> **The big advantage of buying things in a kit is that you get the bits you need** even if you don't know they're the bits you need, and this kit really delivers in this area "

is for putting electronics together, not designing or prototyping them, so it's the perfect companion to soldering kits or prototyping kits (such as the SparkFun Inventor's Kit).

Many people will prefer to build up their toolset bit by bit, and this often makes sense if you're just starting out with something new (you might want to take a look at the SparkFun Beginner's Kit). As you continue in electronics, the Deluxe kit is a good point of reference for extra bits and pieces that you might want to expand your fledgling workshop. However, if you're looking to kit out a workshop with a cost-effective electronics kit, this kit should give you everything you need to work through most basic and intermediate electronics projects. ◻

VERDICT
Everything you need to get started building electronics.